T0269021

**Performing Math**

# Performing Math

• • • • • • • • • • • • • • • • • • • • • • • • • • • • • • • • • • • • • • • • •

## A History of Communication and Anxiety in the American Mathematics Classroom

ANDREW FISS

**R**
**Rutgers University Press**
New Brunswick, Camden, and Newark, New Jersey, and London

Library of Congress Cataloging-in-Publication Data

Names: Fiss, Andrew, author.
Title: Performing math : a history of communication and anxiety in the American mathematics classroom / Andrew Fiss.
Description: New Brunswick, New Jersey : Rutgers University Press, [2020] | Includes bibliographical references and index.
Identifiers: LCCN 2020008435 | ISBN 9781978820210 (hardcover) | ISBN 9781978820203 (paperback) | ISBN 9781978820227 (epub) | ISBN 9781978820234 (mobi) | ISBN 9781978820241 (pdf)
Subjects: LCSH: Mathematics—Study and teaching (Higher)—United States—History—19th century. | Communication in mathematics—United States—History—19th century. | Math anxiety—United States—History—19th century.
Classification: LCC QA13.F53 2020 | DDC 510.71/073—dc23
LC record available at https://lccn.loc.gov/2020008435
LCCN 2020008435

A British Cataloging-in-Publication record for this book is available from the British Library.

♾ The paper used in this publication meets the requirements of the American National Standard for Information Sciences—Permanence of Paper for Printed Library Materials, ANSI Z39.48-1992.

www.rutgersuniversitypress.org

Manufactured in the United States of America

For Laura, Sebastian, and Toby

# Contents

# Preface

It's really hard. I'm not good at it. I don't like it. I hate it.

Math has a bad reputation. Because it is so common to hear these phrases, we often ignore them. In ignoring them, we might find ourselves echoing them as well. People who would never say they hated something else proudly say they hate math all the time. Maybe, when faced with an unexpected calculation or a tricky exam problem, we find ourselves repeating, "I hate it too." Noticing the trend, I took what might seem like a path away from math. I was a math major in college, and I even spent a summer doing research through a program under the National Science Foundation and the Department of Defense. Before that, I had received perfect scores in math classes in high school, and I had wanted to be an applied mathematician. But at some point in college, I started to notice how people around me talked about math. When my friends said they were studying music, or English, or engineering, or chemistry, other people affirmed their choices, talking about how wonderful it was. When I said I was studying math, I often heard, "Oh! I hate that!" It happened so often and with such force that I decided I wanted to study that reaction. I was surprised to find, in fact, that I wanted to study the hatred of math even more than I wanted to study math! I took a trail that led to research and teaching in technical communication, which led me to write a book about how we talk about math, specifically about recognizing how learning math is like putting on a show.

It is important to investigate how we talk about math precisely because math ability is often constructed as natural. When I tried out for the middle school math team, the grown-ups around me talked about how they were not so talented, how they "reached a wall" with algebra or geometry or calculus. It was memorable for me because it was the first time I heard so many adults talking about feeling like failures. That moment was not an amazing success for me either: I was not quick with calculations, I had a hesitation in my speech that was worse in arithmetic drills, and I did not make the team. I was named an alternate, though, perhaps because of pure enthusiasm.

Specifically, considerations of "natural" mathematical ability seem arbitrary. I remember hearing about how some advanced mathematics classes could not be offered at all schools. It was surprising to me when so many other people were talking about individual ability, about how they themselves felt like failures, that some math experiences were cut short because of school policy or opportunity. What differences existed among schools, and how did that play out in mathematics opportunities? I wanted to know more about how we talk about math, especially how we talk about learning math. I also knew, even then, that I wanted to suggest an alternative: a grounding of math communication not in terms of ("natural") ability but in terms of groups of people.

Stories of collective effort, however, prove an uneasy substitute for stories of ability. Yes, I did not find math easy. Especially in college, I needed to spend hours in the library, in professors' office hours, and in study groups with other students. Every time I took a new subject, I needed to learn new terminology and new assumptions that built new subfields of math. Real analysis was a particular chore, since it introduced new (epsilon-delta) proofs to reframe the ways infinitesimals and limits built calculus. We had just learned about limits and infinitesimals, and now we were being told that there was something more basic, underlying those concepts. And it was not just written in the language of math; it was written in ancient Greek! It certainly dimmed my enthusiasm. Every semester, about six weeks in, I was on the verge of changing majors. Then I became more comfortable with the new assumptions, with the new terms, and even with ancient Greek, so I kept going. It was even more exciting to succeed after I thought I would fail. But when I told my friends and family about it, I found those sorts of stories were not what they expected. The British educator Heather Mendick has found that male students rarely talk about working hard at math; instead, they are more likely to say that it comes naturally to them. Female students, more likely to tell stories like mine, usually say they never really succeed because they need to work so hard.[1] I already suspected it was a bad idea to frame math success in terms of ability, and it became clear my stories of near failure were confusing too. I gradually found that math success should be framed in terms of communication.

This book's communicative frame builds on others' work, including some of my professors'. I had the good fortune to be a math major in a department where we learned the best proofs were the ones that could be communicated. When we defined how a sequence, $a_n$, converges to a real number, A, we could start by writing "iff for each ε > o there is a positive integer N such that for all n ≥ N, we have | a_n − A | < ε."[2] But we would also have to be able to say and write out "if for all epsilon greater than o there exists a positive integer capital-N such that for all n greater than or equal to capital-N, we have the absolute value of the sequence a at n minus the real number A is less than epsilon." By including a diagram on the blackboard and conversing with others in classrooms or study groups, we would come to a better understanding, one that could ideally be communicated in many different ways. As one of our professors, John McCleary, recently demonstrated in the book *Exercises in (Mathematical) Style*, mathematical proofs can exist in a variety of forms. Math proofs can be symbolic, written out, or purely visual. Their ideas can be explained through imaginary cities, popular card games, fairy tales, or parodies of NPR hosts.[3] Communicative flexibility, we learned, is also not just at the surface level, a matter of the "fluff" surrounding kernels of mathematical "truth." How we choose to express mathematical ideas determines what can be proven just as much as who can participate.

In fact, mathematical frameworks do make possible certain communicative relationships between people. Take the example of cryptography (the mathematical theory of code-making and code-breaking). It is clear that specific instances of communication are at the heart of the mathematical subfield. As I learned in my introduction to college-level math, cryptography is fundamentally about the transmission of a message between person A (Alice) and person B (Bob). The successful transmission depends on agreement between Alice and Bob: for instance, transmitting a public key (the method of encryption) before a message can be sent. In that case, Alice needs to send Bob her "public key" so that he can encrypt a message back to her: for example, "Hello Alice!" into "6EB69570 08E03CE4." Alice then decrypts the message using her "private key," the method she derived from her public key but kept secret. That method helps her efficiently move from "6EB69570 08E03CE4" back to "Hello Alice!" again. The technique assumes that there is no efficient way of moving between the public and private keys because certain math calculations take a notoriously long time, even on computers. In fact, today's computers "talk" to each other using such encryption very often: it is at the base of internet security standards.[4] Appreciating math communication means, in part, seeing possible relationships between messages and calculations. It means learning and teaching about encryption keys and systems, how certain assumptions about math build the possibilities for certain human relations.

Access to math communication, however, is not universal. As Sara Hottinger shares in her book *Inventing the Mathematician*, the constraints around math expression discouraged her from becoming a mathematician. A talented math student, she anticipated "reaching a wall," failing in summer research programs, the GRE subject test, and applications to graduate programs. She became interested in what she calls "our cultural understanding of mathematics" because she wanted to explain why her math stories did not fit, why she felt compelled to pursue an interdisciplinary humanities PhD instead of one in math.[5] When I read her book, it sounded very familiar to me. Though I was accepted to graduate programs in math, I felt uncomfortable, awkward, and stuck when I thought about actually going to them. I looked forward to the sense of community in an interdisciplinary humanities program. Hottinger encourages us to see these psychosomatic symptoms as part of broader cultural constructions of math. She links her experiences to the work of Mendick and Valerie Walkerdine, who explain the *leaky pipeline* (the attrition of high-achieving math students) by saying that it is difficult for female students to reconcile math success with their identities as women.[6] My experiences, too, could be one instance in the leaky pipeline of minority students. Maybe, on some level, I could not see my math success as "normal" in our current educational climate. Such stories constitute a problem because it is not just a matter of individual careers—our choices exist within broader patterns. Though there have been calls for math education to be reformed, it is still the case that there is a steady downward trend in the percentage of American women and minority students pursuing math in high school, in college, in graduate school, and in careers. In fact, there is clear evidence that the "achievement gap" between white and nonwhite students has been stabilizing or even getting worse over time, from the 1980s to the late 2000s.[7] It is clear that we need a new story.

This book is my argument that math education should be reframed, showing how math education is communication-based and how studying and communicating about math involves a considerable amount of theatrical performance. It is not a book about theater, though it sometimes uses dramatic terms to guide us through a particular stage and help us understand a particular community of players. Much of the story is about nineteenth-century U.S. American colleges (hereafter referred to only as American colleges) around the time of the Civil War because they set the scene for our current educational paradigm in math. The "community of players" are math students and professors experimenting with constructing the blackboard, writing textbooks, burning textbooks, and generally making possible the academic subjects, educational rationales, and teaching techniques that seem unremarkable today. Although I focus on past players, their experiences should uncover assumptions about learning math today. While stories of nineteenth-century American colleges do allow us to look closely at massive historical changes, they are important in part because they present different stories about learning math now.

## Why "Performing Math"?

Until recently, my work has had little to do with performances and shows. There is not much theater in the history of science, and my move to technical communication has been gradual. In fact, there has not been much theater in technical communication either; only recently, there have been movements about what scientists can learn from actors about communicating.[8] Though this book is not organized around teaching tips, it similarly attempts to move us toward envisioning the integration of the arts with STEM communication by looking at the relationship between performance and math.

In doing so, I try not to romanticize math, keeping in mind the strong, negative reactions to it. Though the history of science arguably had its American origins in criticisms of the atomic bomb, much historical work tends to promote scientists and scientific work today. Technical communication works similarly. In fact, some STEM communicators view their field as promoting "scientific literacy," bringing the light of science to the "uneducated" masses.[9] There have been some good books arguing against those ideas, pointing out their elitism. Still, it is rare to find books in technical communication, as in the history of science, that do not take scientists and engineers to be unadulterated heroes. As some practitioners say, why would we encourage the "knocking" of science when our jobs literally depend on its cultural worth? In other words, where would the history of science be without science; where would technical communication be without technologists? This book is therefore a unique challenge, an opportunity to write about math communication when not all the relevant sources even like math.

Some of the challenge of this book does come from complexity. Recently, I had the wonderful opportunity to pitch this book in an Alda communications workshop, one founded on the idea that improv exercises provide the best way to get scientists to talk about their work. My pitch was not the success I had imagined. I mentioned my career path, how it has taken me from field to field in an attempt to write about the question, Why do so many people say they hate math? I talked about my previous interest in math textbooks and how a West Point doodle changed my life. When I was partway through, to the point when I was articulating the main argument of this book, the facilitator stood up. "Your message," he said, "is: I can make math class fun! That's it. Nothing else. Now go to the back of the room and think about what you've done." I was stunned. I had never been sent to the back of the room, ever. Despite the reactions of the other participants (who kindly did try to stand up for me), I think the facilitator was right—at least, in part. This book would be much more usual if my message was "math communication can make math class fun, that's it." But that cannot be the argument. The stories I have collected show people grappling with complex times, places, and personalities—not always fun and enthusiasm—certainly not always for math. Historical evidence aside, there is too much variation in reactions to math today, everything from "math is fun!" to "I hate math!" This book explores a space between math enthusiasm and the hatred of math.

I cannot be so unbiased about performance, especially theatrical performance. When I was trying out for the math team, I was also trying out for the school play. I was the villain in a murder mystery, and it gave me the opportunity to be impulsive, sinister, and mainly just really loud—totally different from my real life. Then I started acting in community theater, something I loved, especially when I got the rest of my family involved. Actually, I was raised by theater enthusiasts. My parents met in the Musical Comedy Society of City College of New York (CCNY). My mother took voice lessons in Manhattan and went to a class or two at the Herbert Berghof Studio, and my father began a theater-themed radio show for CCNY Student Media. My sister and I grew up around the memorabilia of their theatrical lives—black-and-white photos of my mother playing Sally Bowles, Maria, and other leading characters; dusty audio-editing equipment; my parents' extensive record collection; and even a few recordings of the radio show. By the time I can remember, my father's idea had grown into a career, developing into a midmarket radio program focused on local community events, especially performances. Because I grew up around theaters, I had a good idea of the sorts of jobs that went into productions, how it was much more than the actors on stage. This book's focus on "performing math" therefore grew out of that background.

In my own experience, learning math and performing occurred side by side. Consistently working toward a career in applied mathematics, I also continued to act throughout school, and I joined the tech crew—setting up the lights; managing the soundboard; and operating the flies. While a math major in college, I was involved in two theater societies. I helped with a musical, providing some help with makeup, costumes, and props. I acted in a political drama of military life. In a surprising twist, I acted in the only show sponsored by the math department: Tom Stoppard's 1993 play *Arcadia*, about a young female mathematician circa 1810 and the "present-day" historians and mathematicians who study her house and family. When I stopped pursuing math in graduate school, I also stopped being involved in theater, through no conscious connection then.

After teaching at three universities and researching the hatred of math at nearly all of them, I am not so surprised by the reactions to math I encounter; lately, I have been more surprised about how my students feel about theatrical performance. All my students need to encounter math at some level. Requiring math is ubiquitous in high schools and colleges throughout the country, though it might seem, as the writer Andrew Hacker puts it, "a harsh and senseless hurdle." Theater is not considered so necessary—though perhaps the choice is surprising. Poking fun at the hyperbolic rationales surrounding STEM, Hacker proposes PATH instead: an acronym for the power of "Philosophy, Art, Theology, History" or for "Poetry, Anthropology, Theater, Humanities." "*We are falling behind our competitors in PATH pursuits,*" Hacker booms (facetiously). "*If our nation is to retain its moral and cultural stature, we must underwrite a million more careers in PATH spheres every year. If we do not, we may continue to lead in affluence, but we will decline as a civilization.*"[10] As Hacker's half-hearted jokes make clear, there are not so many incentives for requiring PATH today. Though my students' familiarity with math remains strong, they encounter theater less and less. Very few of them have acted, hung lights, painted sets, or otherwise helped put on a show. Almost none of them have ever read a play or seen a production. In fact, almost none of them have even been inside any performing space! I find I need to write this book not just because of the hatred of math but also because of the neglect of theater.

Even with all these interlinked experiences, the idea of "performing math" was somewhat unexpected and came from a surprising discovery—or maybe I should say realization—that I had unique access to a math play from 1886! *The Mathematikado*, written, produced, and performed by Vassar students, was a send-up of Gilbert and Sullivan's opera *Mikado* but rewritten about math class. Literary critic Laura Kasson Fiss had bought herself the libretto at a used bookstore because she thought it nicely combined (and made fun of) our joint interests. Then it sat on our shelves (taken down time and again if we needed a laugh) for about seven years. One day, in research positions at our alma mater, we realized that it fit with the West Point doodle, the math diaries, campus traditions, and the recent scholarship about play and performance. Plus, we quickly realized, though writers and archivists had compiled clippings about *The Mathematikado* from historical newspapers, no one else had a full record of the play.[11] It was the perfect opportunity.

*The Mathematikado* libretto quickly showed how a performative approach to math communication interested people beyond our scholarly subfields. In 2015, I began presenting about *The Mathematikado* to other historians of science. It went so well that I was invited back the next year as part of the featured roundtable about performing science. It coincided with a professional meeting for academics interested in the integration of the sciences and the arts, and "performing science" allowed the coordination of programs among historians, artists, activists, and literary critics. Then the next year, colleagues at the British Science Association decided to sponsor Laura and me to give a *Mathematikado* lecture-performance at their British Science Festival! With a little help from our (British) friends, we organized the singing of four math songs and together gave a lecture explaining the significance of the document. It nicely appealed to STEM aficionados as well as students, journalists, and Gilbert and Sullivan enthusiasts. There was even a little media coverage because it seems the British public was curious about what Americans had done to their cultural treasure.[12] (Answer: They made it about math!) It encouraged interest in performance, history, math, and studenthood, and it began to show how those areas should be considered together.

The project generated so much interest because it was not just about any historical students performing math—it was about women performing math. *The Mathematikado* proved to be a spectacle in the 1880s because educated women were still under a tremendous deal of scrutiny. It was widely touted that women's college education was an "experiment," one that might end in failure. Today, *The Mathematikado* remains a spectacle because there is still a tremendous gender disparity in math specifically. Though many have tried to explain the phenomenon through significant sociohistorical analyses or idealized understandings of the brain or biology, the low numbers of women in math have remained a serious concern.[13] The play can provide hope for a different future by presenting a different story of the past, one where women can do math and have fun while doing it.

Despite the power of that case, *Performing Math* has to incorporate *The Mathematikado* into broader arguments about math communication and the role of performance. For one thing, though *The Mathematikado* seems to be about STEM enthusiasm, it should be understood within the broader dynamics of college students expressing their hatred of math. After all, when young men at Harvard or the Stevens Institute of Technology reported on the production in the 1880s, they thought that was what the Vassar students were doing. They made connections to their own campus traditions, ones that centered on the destruction of school property, even though nothing was explicitly destroyed in *The Mathematikado*.[14] In fact, as in chapter 4, student reporters at other universities recognized the Vassar performers as college students because of the play's supposed commentary on the hatred of math. Therefore, even as early as the 1880s, arguments existed that linked *The Mathematikado* to other artifacts of math communication: doodles, diaries, campus traditions, and other plays, as well as blackboards, classrooms, exams, and textbooks. The frame of "performing math" still does connect to questions of the hatred of math, as well as many other potential reactions. It allows a way to incorporate various student activities of math communication, including but not limited to theatrical performing.

**Performing Math**

# Introduction

• • • • • • • • • • • • • • • • • • • • • •

*Performing Math* focuses on the historical development of expectations for math communication—and the conversations about math hatred and math anxiety that occurred in response. Acknowledging the importance of nineteenth-century American colleges for the establishment of various widespread educational frameworks, this book analyzes foundational tools and techniques of math communication: the textbooks that supported reading aloud, the burnings that mimicked pedagogical speech, the blackboards that accompanied oral presentations, the plays that proclaimed performers' identities as math students, and the written tests that redefined "student performance." Math communication and math anxiety did go hand in hand, too, as new rules for oral communication at the blackboard inspired student revolt and as frameworks for testing student performance shaded into records of performance anxiety.

Math students are the main focus of this book. *Performing Math* had its origins in two archival discoveries: scripts from student-generated math plays and printed programs from college traditions of burning math textbooks. Both types of texts exist in dozens of American educational archives today, and their prevalence shows how past students have represented their relationship to their math classes as performance-oriented, even theatrical. Whether institutionally sanctioned or not, these documents—and the events they proclaimed—had their heyday in 1840s–1890s colleges, specifically where professors and alumni were most active in creating new, American expectations for learning mathematics.[1]

Within the historical context, student plays and even textbook burnings were more than instances of resistance. These events provided rearticulations of pedagogical justifications, students' attempts to assert their own ways of speaking and writing about mathematics.

This book takes performance at the center of a communication/anxiety diode. *Performing Math* suggests we acknowledge how learning math has been about certain rules of performing: how to read, how to speak, how to write, even how to rehearse and how to put on actual plays. Given these components, math anxiety appears a kind of stage fright, a reluctance to follow the conventions of oral and written communication. Recovering the performative dimensions of math education, this book urges readers to reconsider learning math as communication-based, as about embodied people and not disconnected minds.

Academic interest in performance has surged in recent years, especially through discussions of gender. Judith Butler famously argued that gender expression and gender identity are the same, that "there is no gender identity behind the expressions of gender; that identity is performatively constituted by the very 'expressions' that are said to be its results."[2] Looking beyond drag traditions, Butler finds echoes of gender "impersonation" in the work of theorists, writers, psychologists, and biologists, especially those who argue for the importance of some psychologically grounded (or biologically grounded) gender identity. Butler instead argues there are no core thoughts/parts that determine gender, that even people who identify as women impersonate womanhood by acting in certain ways, men impersonate manhood, and so on. Following Butler, many scholars have reiterated the need to look beyond identity categories that are supposed to have some prior reality; instead, they urge, there must be a focus on social performance. For the purposes of this book, such perspectives begin to indicate the gendered complexities of textbook funerals and math plays, as well as the more mundane ways of acting like a math student: reading textbooks, speaking at the blackboard, and taking written tests.

*Performance* has become a useful term because it flags approaches not only to gender but also to scientific knowledge. As historian Delphine Gardey recently observed, Butler's views developed at the same time—and in dialogue with—perspectives from the Sociology of Scientific Knowledge (SSK).[3] Before SSK, historian Thomas Kuhn had suggested a new reading of the development of science, beyond the stories told in textbooks: that science changed through remarkable moments ("revolutions") but generally remained consistent. In fact, the consistency of science, for Kuhn, could not just be explained through textbooks and other repositories for logical rules; it needed to be about some unspoken skills, something like a pianist's sense of fingering or a craftsman's sense of their craft.[4] Proponents of SSK similarly explored these ways that science was much like other parts of human life, through treating laboratories as sites of "foreign" culture or considering scientific disciplines as networks of trust.[5] Though math perhaps has even longer periods of consistency than the other sciences do, it can be considered through similar frameworks.[6] The rules for speaking at the blackboard, for instance, are usually tacit, learned through observation and imitation and not through stated directives.[7] Such situations in math require analysis not only because of the ways that they generate knowledge but also because they are so rarely talked about.

Acknowledging these traditions from gender studies and science studies, *Performing Math* encourages us to consider the links between math and performance. For generations of Americans, theatrical displays about math (textbook burials and actual plays) were ways of reiterating their identity as students. Many others found their mathematical knowledge needed to be put on display through interacting with the blackboard, taking written tests, or even reading. By recovering such moments, we notice and analyze common expressions, such as the command that students need to "perform" calculations.[8] This book centrally asks, When are students said to "perform" math, and what are the implications?

In answering the previous question, *Performing Math* looks back to nineteenth-century colleges. As many historians have noted, much of our current educational system emerged around the time of the Civil War.[9] Therefore, telling stories of the past can help expose some of the tacit expectations of American education, especially surrounding the math classroom. Analyzing moments of change and disruption, *Performing Math* begins to tell us where these rules for math communication come from and why they should matter. It emerges that when faced with new expectations for their behavior, many American students began to talk about hating mathematics. Through these stories, a history of math communication should be a history of math anxiety as well.

## Math Communication through Technical Communication

*Performing Math* uniquely combines approaches from the history of math education and an expanded view of technical communication. Technical communication, like math, is both very broad and very specific. Broadly, it refers to any messages (verbal, nonverbal, visual, physical) that include terminology from specific fields or groups of people. An example is a recommendation report that an IT department compiles, suggesting whether their workplace should go paperless and require all staff to carry tablets instead. Though written for an audience of supervisors who likely are not IT people, the report might include some terminology more familiar to technologists: the BYOD (bring-your-own-device) model and names of tablet manufacturers, tablet providers, and records management systems.[10] Though some might argue that all technical information needs to be replaced with common words and phrases, that is not a wise approach. Such terminology forms the backbone of career training and education and therefore serves as a marker of expertise. Without it, the supervisors might suspect that their IT people do not know how to do their jobs. Another problem with the goal of using "common words" is that it is impossible to say what is "common" for all people, even all people in a given workplace. Technical communication has to occur.

Thinking beyond the one example, too, it becomes clear how technical communication touches many areas. Technical communication means the writing of reports, as well as presentations, marketing materials, instructions, and even résumés. In fact, technical communication is often equated with workplace communication, particularly related to the STEM fields (science, technology, engineering, and math). That said, there is a narrower definition even more specific than STEM workplace communication. Under that definition, technical communication means the communication that happens at engineering firms or other workplaces involving engineers. That does mean a heavy emphasis on math, as well as the $S$ (science) and the $T$ (technology). But some technical communicators do see their field as distinct, as touching mainly engineering.[11] Despite the uses of math in engineering, the link between math and technical communication is just starting to be recognized. This book aims to put math in technical communication, to put the $M$ back in STEM communication.

As we will find, recognizing the $M$ in STEM communication requires putting an $A$ there too. The arts have not always been recognized under technical communication, even beyond the narrow definition. If the goal of technical communication is the training of efficient speakers and writers for STEM workplaces, then it involves developing what the entrepreneur Carolyn Miller criticizes as "a series of maneuvers for staying out of the way" of STEM facts. Under Miller's critical view, technical communicators might be described as conveying others' "discoveries" with as little personality as possible: using stylistic exercises, unemotional tone, and an understanding of psychosocial "reading level" for the "right" vocabulary. Not surprisingly, mathematical proofs are (incorrectly) offered as the epitome of such an approach. As Miller points out, these views of technical communication require a systematic misunderstanding of STEM work. Communication is not irrelevant. Many scientists, engineers, and yes, even mathematicians work in groups, and collaboration requires effective communication. Even working independently, STEM professionals do need to speak and write to others about their work, to secure grants, publications, or at least positions. (After all, could a mathematician be recognized as a mathematician without a job that said so?) An education in the arts, Miller argues, gives insight into the human collaboration and human creation inherent in STEM work. Arts subjects teach ways of understanding "how to belong to a community" and how communities make and thrive.[12]

Moreover, arts approaches provide a way of glimpsing human creativity, which many mathematicians see as central to their proofs.

This book builds on approaches that say the arts can not only help technical communication but also be considered under a framework of technical communication. Yes, machines are used in a variety of arts settings. Scene shops use scroll saws, as do some sculpture studios. Writers use computers, as do set designers and directors. Many folk artists, crafters, makers, and costume designers use sewing machines. Any manuals for these devices can be considered under a narrow definition of technical communication because the production of these machines (and their manuals) involves engineers. That is not equivalent to the consideration of the arts under technical communication, however. Communication scholar Katherine Durack, writing in the 1990s, found that her colleagues drew an arbitrary line between *technical* sewing texts and *nontechnical* ones, even when considering the sewing machine. Though manuals were a clear case of technical communication, apparently sewing patterns were not. Inspired by these conversations, Durack did a systematic analysis and found that such an understanding of "technical" depended on three main assumptions: that militaries and laboratories are the only major sites of invention, that the workplace is outside the home, and that all jobs are equal to each other. In her central case, sewing patterns are rarely created in labs, many are made in the home, and sewing has a legacy of being associated with "housewifery," which has been historically undervalued. Sewing patterns do not fit the assumptions around technical communication. Durack points out that these lines between *technical* and *nontechnical* are arbitrary and in fact perpetuate social inequalities regarding what can be counted as work and what cannot.[13] Following her analyses, I incorporate the arts into technical communication, not just as a helping hand but as a core. Performance and math should both be considered under technical communication.

## The Distrust of Math in Technical Communication and Beyond

In technical communication and beyond, however, math is rarely presented positively. Today, as over the last two hundred years, books, articles, movies, and even toys are more likely to talk about disgust with mathematics. Discrimination, quantification, obfuscation, and boredom have all drawn attention, creating a small industry out of the trope of the boring math class. The 1980s fiasco over a doll that proclaimed "Math class is tough!" continues to be in the news.[14] Standardization, especially the reduction of students and workers to numbers, receives media attention every year in the United States, especially around SAT/ACT season. In fact, so many people experience math-related "performance anxiety" that psychologists have named the condition *math anxiety*—and found it all over the world. Though there is debate about treatment, from parent involvement and teacher attitude to relaxation and visualization techniques, demonstrations of math anxiety have relied (ironically) on highly quantitative methods, leading some researchers (who were themselves "good" at math) to believe that math anxiety meant some sort of "natural" inability.[15] The errors in judgment mirror the larger social debates. Though it is common for many Americans to express disgust with math, anxiety around math, and even hatred of math, others try to excuse such statements as equivalent to being "bad" at mathematics. Because of such moral judgments (being "good" or "bad" at math), debates about math performance and math communication continue to appear because they have very high stakes. This book presents a more nuanced and hopefully more positive view while still acknowledging math's poor reputation.

The distrust of math appears in books in technical communication, as well as popular accounts. Cathy O'Neil, in her exposé about algorithms' perpetuation of social inequalities, *Weapons of Math Destruction*, uses her extensive background to uncover certain big data algorithms used in insurance, advertising, education, and policing, pointing out the ways some math can be opaque, unregulated, and (un)scalable, amplifying existing social biases and extending racism, discrimination, and social harm.[16] In the field of technical communication, Beth Flynn's 1995 article "Feminism and Scientism" remains a core critique of highly quantitative methods. Noticing the uses of quantification in preserving existing power structures, twenty years before O'Neil's book, Flynn used her literary and feminist research skills to understand the assumptions about mathematical objectivity and neutrality that perpetuated these cultural impressions. Observing how technical communicators have been characterized as "sad women in the basement" at her university (literally, where I am writing now), she reviews the popular and scholarly literature to analyze what it means for communication/writing to be "feminized." Her analysis ultimately focused on how technical communicators and compositionists attempted to get out of "the basement" through identifying with fields that have more cultural power, such as mathematics. As a result, Flynn calls for scholars in composition and technical communication to "embrace . . . discourses of resistance," ways to neutralize and disrupt the uneasy partnerships between quantitative disciplines and technical communication.[17] O'Neil, more recently, talks about her involvement in the Occupy Movement as a way to protest math's role in social injustice.

Even scholars arguing against the evils of quantification tend to perpetuate a distrust of mathematics. Davida Charney, in a public address against Flynn's article, argued for a nuanced view of mathematical methods, but she spent just as much time characterizing an antimath view as arguing for complexity and reconciliation. She repeatedly invoked Theodore Porter's 1995 book *Trust in Numbers* for its historical cases showing that it is possible to be promath and antielitist and that it is possible to adopt mathematical methods for the purposes of facilitating communication, especially among international groups that do not share linguistic fluency otherwise. Yet Charney also spent much of her time parroting what she saw as views of the "critics." One particularly long section ended, "no one likes the way scientists seem to privilege numbers and disparage words—the way numerical and graphic evidence is treated as clean, precise, and solid . . . ways for scientists to avoid interpretation, eliminate the human element of subjectivity that supposedly contaminates the study of individual cases, and go on misrepresenting the world as manageable, fully determinate, and reducible to clear and accurate formulas."[18] Though Charney likely made such arguments only to refute them, children of Holocaust survivors saw their experiences in her words. Later articles and books considered the ethical implications of technical communication with renewed vigor.[19]

In fact, math and communication are often presented as oppositional. Though academic C. P. Snow characterized the "two cultures" of 1950s British education as "literary intellectuals" and "physical scientists," it has become common to talk about them as the humanities/arts versus the sciences. Though our understanding of the central groups has changed, the problem of the "two cultures" remains a problem of understanding: that these two groups of highly educated people cannot seem to communicate with each other. In making his point initially, Snow began with an apocryphal anecdote about an Oxford professor visiting Cambridge. Seated at dinner with the faculty, the visitor tries to talk to the men on either side. He tries to engage one, who grunts in response. Then he turns and tries to make conversation with the other, who grunts. Then the two talk over his head, trying to figure out what the visitor just said. Finally, the president calls out not to worry: "Oh, those are mathematicians! We never talk to *them.*"[20] These insular groups of math people and nonmath people worried Snow—and continue to worry many people today—because of the recognition that many of the world's problems can only be solved through people with different skills working together. The specific educational backgrounds have not mattered much, which accounts for Snow's quick slippage from "mathematicians" to "physical scientists," but the two cultures remain a way for us to talk about how groups of highly educated people cannot seem to understand (or stand) each other. Today, a fundamental misunderstanding remains surrounding mathematics. In fact, even the "third culture" of STEM communicators, supposedly bridging the two cultures, is cautioned that their writing about math can put off their readers.[21] Even for readers/viewers of STEM media, math is supposed to cause too many misunderstandings.

Popular portrayals of mathematicians similarly emphasize the opposition between math and communication. The 1997 release of *Good Will Hunting* emphasized the transformative power of math, how a Boston janitor was able to recognize his talents and pursue other jobs through his conversations with a therapist and a decorated math professor. Though Will Hunting's transformation depends on talking with others, his character is fundamentally terse, and his math talents show up in ways that necessitate only a blackboard, which he accesses at night, alone.[22] Movies from the early 2000s, such as the 2001 *A Beautiful Mind* and the 2005 *Proof*, continued to portray mathematicians as awkward communicators. In such cases, forms of mental illness, which perhaps heighten characters' math abilities, also make them difficult to understand.[23] The 2016 *Hidden Figures* proves a notable exception for its portrayal of Katherine Johnson and her talented NASA colleagues. However, some debate remains about how much *Hidden Figures* is "really" about math because the characters are so good at expressing themselves.[24] In short, math and communication continue to be seen as at odds.

## Why Study Math Communication?

I point out negative images of math—and the assumed opposition between math and communication—in order to indicate the lack of recognition for "math communication." In fact, following the articles, books, and movies mentioned previously, it would seem "math communication" could not exist, that there would be an internal contradiction. In fact, as in the "third culture," popular writers, even popular STEM writers, worry that just mentioning math would put off readers. Moreover, most of the previous accounts relate to contemporary issues; there are very few books that use historical reflection to build a case for math communication. George Sarton, the chemist and historian who helped popularize the history of science in America, characterized historical math writing as *secreta secretorum*: "If the history of science is a secret history, then the history of mathematics is doubly secret, a secret within a secret, for the growth of mathematics is unknown not only to the general public, but even to scientific workers."[25] Math communication is similarly limited, "a secret within a secret," and this book in part seeks to explain why.

The hatred of math is not a recent phenomenon but has a history at least as old as the country. As early as 1843, some Yale sophomores participated in the "Burial of Euclid," burying an effigy of the ancient Greek mathematician Euclid while saying generations of American students had already been observing "his" funerals. Why bury someone long dead? As the Yale students put it, it showed how much they hated Euclidean math, how they were glad to be rid of it.[26] Given that most American colleges had versions of such a tradition, the enduring popularity of math requirements indicates that there must be some social advantage that comes with mathematical knowledge. As the 2016 book *The Math Myth* points out, most college degrees still require some level of math (usually calculus), even though there are not many cases where college-educated professionals would need it on the job. Math has been and continues to be a site of privilege, but one that is not so easily justified. In fact, appeals to tradition, to career readiness, to "natural" talent, even to hard work do not satisfy today's critics.[27] Math communication, how we talk about math, deserves further attention. Like introductions to theater, perhaps our introduction to math communication needs to explore origins and histories in order to understand how Americans' relationships to math came to be. In other words, to explain the present and imagine the future, we need to develop what a 1990s theater textbook calls "a way of seeing," one that incorporates the histories of Americans performing mathematics.[28]

Following such approaches, this book begins by recognizing that math communication is a product of particular histories and historical contexts. It does depend on a variety of cultural factors that are seen as somewhat stable: assumptions about who counts as an expert, where to find sites of education, and even what constitutes math itself. Yet as we will see, even many of these views have their roots in Civil War–era practices. The communication of mathematical ideas, to a variety of audiences in a variety of ways, has been a product of specific debates about gender, race, and power in the United States. As a result, it is very difficult to understand today's attempts at math communication without understanding the social and cultural practices that supported their development. In fact, the assumptions about stability do not necessarily mean complete stagnation. Math communication has experienced change—from influences internal and external to math and education. Our framing in terms of math communication can be an asset here because it avoids deep philosophical debates about how math can change. I propose we start with changes in math communication, specifically in Americans performing mathematics. Math has maintained social relevance for millions of students (whether they like it or not), and so its communication deserves our attention.

Paying attention to math communication also justifies my use of the term *math*. Though nineteenth-century Americans more often used the term *mathematics*, the abbreviation flags more recent developments, especially in "math anxiety." Also, I should point out that its usage is not entirely ahistorical. As the *Oxford English Dictionary* notes, the usage of *math* as an abbreviation dates to at least the 1840s, particularly in the United States. As early as 1847, "math" was used in the context of college remembrance, and American newspapers began to use it as common slang beginning in the 1870s–1890s. While the phrase "you do the math" seems firmly twentieth century, the abbreviation "math" does flag the historical time that I am discussing as well.[29]

This book therefore emphasizes historical methods in addition to the perspectives of technical communication. It focuses on college life around the time of the Civil War because of the ways that today's academic subjects, educational rationales, and teaching techniques developed then. I have used the same sorts of sources that other historians do: textbooks, college archives, personal papers, newspapers, and public addresses. I began collecting material for this project when I was a graduate student in a department of history and philosophy of science. At the time, I was interested in how changes in math textbooks mirrored larger social trends in nineteenth-century American history (industrialization, militarization, educational expansion, and educational standardization). From there, the project grew.

Had these documents been my only source, I would not have been able to justify writing beyond educational history. But when I was researching math textbooks at the West Point archive, I found an artifact usually ignored by historians: a doodle. It was about the size of an index card, wedged between the pages of a book, and it clearly showed an 1860s student, "Cadet Dahlgren," being attacked by the "demons of geometry."[30] It seemed like an amazing discovery. It was so much like the late twentieth-century mantra "Math class is tough" or the twenty-first-century responses "Oh! I hate math!" But it was an insert, perhaps from the nineteenth-century diarist himself. Even though I was interested in histories that incorporated student perspectives, I had not seen a historical doodle incorporated into any other book, let alone any other dissertation. It did not appear in mine either. Though I fought to keep it in, it asked too many questions and left too many paths open. In fact, in part because of the doodle, I left the history of science for technical communication, where I could find some possible research paths through omission, what had not been said about math communication and the role of responses.

In fact, the doodle was the beginning of a large work of recovery. Student doodles are difficult to find, but there are a lot of similar materials in used bookstores, archives, college grounds, and performance spaces. Plays, diaries, exhibits, and school traditions (including textbook burials) allow access to student perspectives, and along with doodles, they provide diverse stories about the potentials and pitfalls of math then and now. *Performing Math* takes advantage of the different access point, a different way of understanding math communication. Though many have followed the usual path in focusing on the importance of textbooks, textbooks do not capture the activities of students—playing, joking, talking, doodling, attending, and distracting. In math classrooms, these happenings are more than the sum of static textbooks; they constitute the bulk of math communication.

My goal for this book then is to use historical examples to build a sense of learning math that is embodied, creative, complex, and exciting. In doing so, I hope we can understand math communication as performative, even dramatic, and the activities of American math students as far from predetermined. My historical examples span the last two hundred years, though they especially come from the Civil War era because that was the time when our current paradigms of education came to be. This book is more historical than many accounts of the public understandings of STEM, communication, or math education. It recognizes there are other books about math communication (mainly from the 1980s and 1990s), falling into three main camps: linguistic analyses, teachers' perspectives, and self-help manuals about overcoming math anxiety.[31] My historical approach should supplement these, bringing their suggestions together under a larger umbrella.

This book also covers more than traditional historical works. For many historians, the events of a year or two are plenty to fill a book. In taking a much larger scope, this book also follows the activities of thousands of people: professors, teachers, and students (mainly) with some university presidents, military officers, actors, and artists. "Performing math" captures their generalized experiences, though it tries not to lose sight of the quirks of individual lives. American math communication, after all, exists in the annual cycles of the school year. It exists in the cycles of studenthood: the time each individual spends in schools and colleges. It exists in political-social transformation: the times of upheaval that test social expectations. Math is difficult for historians to write about because it is expected to change slowly, to outlive its calculators and computers.[32] Students' math communication is equally difficult because it is expected not to stay around. Doodles are fundamentally ephemeral, as are plays and performances. (No one today can truly capture the experience of seeing a production from 1886.) Yet something of math communication remains, and something of math changes. This book proposes the common thread is in the performative dimensions of learning mathematics.

This book therefore builds on the work of other researchers, similarly interested in exposing the many people who work together toward math education. Supplementing the usual focus on parents, teachers, and students, a group at the Smithsonian Institution has been drawing attention to technicians and tools. Beginning with the recovery of stories about textbook authors, their book *Tools of American Mathematics Teaching* also points out the individuals and companies who made blackboards, overhead projectors, slide rules, protractors, graph paper, models, calculators, and much more. Their goals in writing seem to be to increase our understanding of these objects as having histories/stories and to encourage our drive to preserve them in museums.[33] (After all, many of the objects in their book came from the collections of the National Museum of American History.) Their book usefully incorporates different kinds of mathematical makers, not just teachers and students but the people who made blackboards and protractors. Like museum studies, performance also suggests ways to understand the many people who have shaped the experience of learning math in America.

Further implications follow from the performative focus of this book as well. *Performing Math* does not just prove the argument that math education has performative components, though such connections provide an important thread through the next chapters. More generally, performance allows for the introduction and investigation of broader questions in education studies. First, as indicated in my discussions of math hatred and math anxiety, arguments about performance do have important implications for understanding affective responses to education. Second, performance introduces the "justification problem" in mathematics—meaning, the reason it is studied and why such rationales would matter. Lastly, most broadly, the term flags assumptions not only about theatrical communication but also about a certain outcomes-based framework for business, education, and beyond. (Consider "performance anxiety" or "performance assessment.") In short, a performative focus allows for the investigation of potential reactions to math anxiety, math hatred, and their undergirding frameworks.

With performance, the focus on mathematical makers helps explain why my book emphasizes examples from the late nineteenth century and early twentieth century, roughly from Civil War–era America to World War I–era America. *Tools of American Mathematics Teaching*, after all, has a similar temporal focus, as do many books about the history of American math education. The time period witnessed five great expansions: in the types of schools available (roughly, from colleges, academies, and town schools to common schools, high schools, universities, and other institutions of many kinds), in the mission of higher education (often characterized as a move to the sciences and specialization), in the demographics of students and faculty (as women and members of minority groups became educated at unprecedented rates), in perceptions of the needs of education (through urbanization and industrialization), and in professional organizations (for teachers and specialists of various fields).[34] Other books have clarified the historical period as the time of "emergence" or "the introduction of new practices."[35] It was also a time of professionalization, following from new gender roles, jobs, and ways of being in the nation. Learning math is like putting on a show in part because as (theatrical) performances depend on the work of many people doing many jobs, so too does math education.

## Performing Math in This Book

The combination of performance and math will be clearest in the organization of this book. The chapters oscillate between routine, expected activities of math communication (reading, speaking, and writing) to more disruptive ones (burying textbooks and acting in plays). The analysis of students' explicit performances (say, on stage) will complicate and clarify the expectations for classroom behavior. Such an organization makes tacit understandings explicit and emphasizes how many people over many years have worked together to create the possibilities of learning mathematics.

My first chapter, "How Math Communication Has Started with Reading Aloud," introduces learning math through a usual staple of math education: textbooks. Our current terminology of "recitation sections" alludes to the past practice of reading aloud from math books. In fact, nineteenth-century textbooks were written for the purpose of being recited, embodied like plays or other explicit performances. Chapter 1 therefore allows us to explore how math communication has started with reading aloud. In part because of the importance of recitation, student experience has been defined in the words of textbook authors.

The second chapter, "How Math Communication Has Been Practiced in Prohibited Ways," focuses on the other activities of professors and students, beyond reading aloud. As we shall see, math communication has depended on practice. In the nineteenth century and since, professors have tried to justify their classes, repeating reasons their studies are important. Students have rehearsed similar speeches, though in unexpected ways. Even in campus traditions, even in the destruction of school property, students have incorporated and gone over their professors' rationales, the reasons math should be studied. These student activities, usually ignored as merely disruptive, have had pedagogical value, I argue, as ways of practicing for sanctioned outlets of math communication.

"How Math Anxiety Has Developed from Classroom Tech," chapter 3, provides an opportunity to consider the historical environments of learning math and their implications. Math communication has depended on various mathematical makers, especially those who create the tools that support mathematical messages. Focusing on classroom design and the use of the blackboard, chapter 3 follows certain people who made tools marketed at math students and educators and who defined new rules for communicating with new technologies. The implementation of pedagogical tech, though, often came with unanticipated consequences. Math anxiety, as a form of stage fright, has emerged through student interactions with blackboards.

Math communication also has had more explicit connections to theater, the focus of chapter 4, "How Math Communication Has Been Theatrical." Chapter 4 discusses educational performance traditions and mathematical plays, especially student-generated ones from the late nineteenth century and early twentieth century. These plays have mattered for performers' identities not only as math communicators but also as students. Math has been a way of proclaiming studenthood, and theatrical outlets have served to magnify their message.

Chapter 5, "How Math Anxiety Became about Written Testing," follows the continuing story of math anxiety. In chapter 5, I begin by discussing the origins of math anxiety in the tradition of public examinations: oral exams before large groups of observers from schools and beyond. It explores how the standards of math testing changed from these oral events to written ones over the course of many years and with the assistance of newly minted educational researchers. Trying to control the message, educational researchers have pushed for the spread of written testing, the importance of the quantitative study of students, and new, explicit definitions of *student performance*—all of which intersect in our current paradigm of math anxiety.

In the conclusion, "Math Communication from STEM to STEAM," I consider final thoughts about the connections between math, performance, and education. It briefly explores the ways math has been positioned as an art rather than a science, entertaining the argument that this book might have more in common with art history than the history of science. In doing so, the conclusion offers some specific suggestions about learning math, mainly commenting on recent work about STEAM as a way to build a new vision of math communication and math anxiety.

As I hope my introduction has already made clear, performance and math share many characteristics. They both depend on the work of many people, some of whom are undervalued because their work seems too much like playing. They both depend on processes and systems, though the process is sometimes obscured. They both involve anxiety, as in stage fright or something much like it. They both elicit strong emotional reactions, whether on purpose or not. Their communication frameworks overlap, and their histories intersect. In sum, *Performing Math* exposes how studying and communicating about math involves a considerable amount of theatrical performance.

# 1

# How Math Communication Has Started with Reading Aloud

● ● ● ● ● ● ● ● ● ● ● ● ● ● ● ● ● ● ● ● ● ●

On Saturday, July 31, 1830, forty-five students were dismissed from Yale because they disagreed with the faculty about how to read from their math textbooks. Many had come from preparatory schools and were descended from established families. One, in fact, was the son of a U.S. vice president. The forty-five said they should be expected to read solutions from their textbook, not develop them from the diagrams printed in the book or reproduced on the blackboard. The new policy, they claimed, made their algebra classes "one-half more difficult" than those of previous Yale students, and conic sections (their current class) was just not possible for them.[1]

President Jeremiah Day and his faculty responded with dismissal. Not clarifying if the decision was equivalent to expulsion, they still reinforced their action by circulating pamphlets and sending letters to university presidents, professors, and especially parents. The students, they explained, had broken the rules by "enter[ing] into combination" and banding together as a unified body against the faculty. Furthermore, the faculty did their own quantitative study, which did not find, as they said, "greater burdens" in these students' math classes. Ultimately, though no official expulsion, none of the students could return to Yale, and many found they could not complete a college degree anywhere.[2] The forty-five found they had fundamentally damaged their reputations. Why? They disagreed with their faculty about how to read from their textbooks, about how to communicate about mathematics.

Historians have long been fascinated with the story of the Yale Conic Sections Rebellion of 1830. In his history of middle-class professionalization, Burton Bledstein placed it within the context of emerging youth cultures that signaled social change.[3] Similarly, Henry Shelden considered it a representative anecdote of changing student life.[4] The various chroniclers of Yale University, including Brooks Mather Kelley, Clarence Deming, and William Lathrop Kingsley, have linked the Conic Sections Rebellion to the other Yale protests of the era.[5] More recently, math historians Peggy Kidwell, Amy Ackerberg-Hastings, and David Roberts have found it indicative of the blackboard's pedagogical novelty.[6] Chapter 3 will return to the historiographical framing of the Conic Sections Rebellion, arguing for its importance for our understanding of math anxiety as a form of stage fright. For the purposes of this chapter, the Conic Sections Rebellion flags the importance of establishing standards for reading in American mathematics classrooms.

Math classes then—and since—have featured strong opinions about how best to read textbooks. As many historians of primary schooling have recovered, classes in reading preceded those in math; the famous three Rs—reading, (w)riting, and (a)rithmetic—had a particular order.[7] First, students learned how to read. Then they learned how to write. Finally, they learned arithmetic. Studies in arithmetic took years because of the framework for memorizing rules for specific types of problems, such as the "Rule of Three." There were also extensive units about business and commerce, such as the lengthy tables of the monetary systems of England, France, and different parts of the United States.[8] Throughout, reading was considered the best way to access arithmetic, as foundational for any math studies.

Day, who chastised the participants in the Conic Sections Rebellion, published textbooks that followed the common practice of beginning with advice about how best to read the book. Day's *Algebra*, a central textbook for Yale's forty-five, included a long, five-page preface along those lines. It began with the problems of the existing, British models: too long and detailed for American students; or too short and sketchy for American classrooms. The implication was that American math needed to be homegrown. British books assumed advanced learners or access to a system of tutors and professors. The United States, without the resources of established British universities or the preparatory schools that supported them, needed a new kind of book. Building on the British models, proudly improving them, Day's textbook ended its first sentence with a mission and promise: "accommodated to the method of instruction generally adopted in the American colleges."[9] Such a message, with updated language and perhaps substituting "schools" for "colleges," became common, as did the claim that American schooling necessitated American books. Fundamentally, for Day, the difference in "method of instruction" was precisely a matter of reading.

Reading, for Day, had to do with cultural expectations for instruction. "In the colleges in this country," he wrote, "there is generally put into the hands of a class, a book from which they are expected *of themselves* to acquire the principles of the science to which they are attending; receiving, however, from their instructor, any additional assistance which may be found necessary." In other words, it was distinctly American for math students to be expected to read their textbooks and learn for themselves. "An elementary work for such a purpose ought evidently to contain the explanations which are requisite, to bring the subjects treated of," he continued, "within the comprehension of the body of the class."[10] A successful American math book, for Day, needed to be so thorough that anyone reading it could come to full "comprehension" with minimal involvement of a teacher—or perhaps none at all. With textbooks good enough, "accommodated" enough, American students could just go off and teach themselves.

The role of reading was further clarified in Day's continued assertions that American math textbooks should not be "practical." Yes, he admitted, a lot of people learn without reading. By figuring out what examples fit what situations, workers build careers in occupations related to math: "In this mechanical way, the accountant, the navigator, and the land surveyor, may be qualified for their respective employments, with very little knowledge of the *principles* that lie at the foundation of the calculations which they are to make."[11] But, Day implied, such workers only act like machines and can be easily replaced. The derogatory use of *mechanical* (automatic, machinelike, mindless, and soulless) further emphasized that the earlier "*practical*" was meant as a criticism of learning without reading.

Reading about math was important for Day because he thought it provided a direct link to the mind. As a minister, Day liked to know how to sway congregants, how to persuade them to pursue the right path. The link with math was not about a sort of religious persuasion, at least not directly. Instead, reading about math led to logical thinking, a heightened mind through good reasoning. Not about "practical" skill, "a higher object is proposed, in the case of those who are acquiring a liberal education. The main design," he explained, "should be to call into exercise, to discipline, and to invigorate the powers of the mind. It is the *logic* of the mathematics which constitutes their principal [*sic*] value, as a part of a course of collegiate instruction. The time and attention devoted to them, is for the purpose of forming *sound reasoners*, rather than expert mathematicians."[12] Compared to British counterparts, the American textbooks had to be the right length and have the right tone so that readers could teach themselves the "logic" in mathematics. Such framing did not communicate exactly how to think like a mathematician and certainly not like a math worker. Undergraduate studies emphasized how to be a lawyer, doctor, minister, or educator instead—and how reading would further all those careers.

Day's way of explaining math did indicate the specific opportunities and privileges of the imagined audience. Like the chastised son of the U.S. vice president, American college students had tremendous privilege and advantage at a time when less than 1.5 percent of the eligible U.S. population pursued higher education.[13] They often were not the same men (Day consistently assumed them to be men) who had to worry about how accounting, surveying, or any other business would bring them money. They could spend some of their time reading math textbooks and figuring out how it would lead to a better mind through "exercise" and "discipline." They perhaps did not have enough leisure time to become "expert mathematicians." But they could read a particular kind of textbook, not too long and not too short, not too detailed and not too sketchy. Gathering in classrooms full of white men, they could find out how to be "sound reasoners" by participating in performative, collective read-throughs.

Still, students were not entirely empowered in systems of mental discipline. After all, Day and his compatriots assumed student minds needed "discipline," even beyond moments of student rebellion. Partially, Day's view had to do with students' ages. At Yale, students' ages ranged widely between young teenagers and thirtysomethings, though the curriculum skewed young. Before high schools became common almost a century later, American colleges catered to a group that some historians (anachronistically) call *adolescents*.[14] As the term makes clear, college students were assumed to be immature and flighty, certainly unable to think clearly. At the time, their reading needed to be carefully monitored and controlled so that it exerted the best, civilizing influence on them. That way, they could become religious and civic leaders, persuading others through their example, their words, and their "sound" logic. Reading, especially about math, seemed the ideal way for educators like Day to instill American values and exert control.

Reading provided a major way of limiting the audience for math textbooks as well. Day's *Algebra* and the other books of his series imagine an audience beyond Yale, though only "in the American colleges." Unlike some of those British textbooks he criticized, Day's did not assume a general readership, not because of expectations that general readers could not stomach mathematics. Instead, it was a nod to the way that school policy constrained the audience for math textbooks. At the time, most schools and colleges had admission criteria that required all students to be white. The admission of African American students caused a stir in a country that still supported slavery. Similarly, Native American students were rarely incorporated without public outcry and also (often violent) missionary zeal from the educators.[15] As many schools and colleges held admission policies based on race, many required that students be male, though some noted how women could be educated for home and family, including family businesses.[16] Yale students were likely to be white, male, and from relatively well-to-do families. Within these categories, there was the most variation in social class, though even that was fairly limited. In short, Day's "American colleges," imagining from Yale outward, already incorporated implicit links among reading, math textbooks, and expectations regarding race, class, and gender. The Conic Sections Rebellion led to serious consequences because it was not how civilized, white men were supposed to behave, especially if they were truly disciplined, logical thinkers.

It was—and continues to be—important to recognize expectations for reading at the beginning of learning mathematics. The Conic Sections Rebellion included a disagreement about what part of the textbook was supposed to be read during the class time called "recitation." Though our modern sense of the word might lead us to suspect recited texts were supposed to be memorized, "recitation" here meant reading aloud, in front of other people. Before their peers and tutors, the students were expected to read the words in the explanations included in the textbook, but their professors and textbooks' authors encouraged them to "read" the diagrams instead.

More broadly than the disagreement about privileging alphabetic strings or visual diagrams, the Conic Sections Rebellion presaged the more complex expansion of reading abilities to quantitative literacy (or QL). According to technical communication researchers Crystal Colombini and Sue Hum, quantitative literacy includes assumptions about exploration, translation, visualization, and expression, touching on information literacy, visual literacy, and tech literacy, among others.[17] In other words, reading about math—or as they put it, reading "about data"—involves alphabetic reading ability, along with the ability to "construct meaning" from images and information, usually accessed through digital technologies.[18] In the Conic Sections Rebellion, Yale students claimed not to have been prepared for such an expanded notion of reading.

Analyzing the role of reading in math classrooms begins to indicate the performative dimensions of math communication. Teachers, professors, and textbook authors have had certain expectations about students' minds: how to access them, how to inspire them, and how to discipline them. By setting expectations for reading, past educators created a sense of order as an introduction to learning math: an appeal to reason and logic as a way to exert control. As in the continued use of the term *quantitative reasoning* (or QR), discussions of reading and math still discuss the project of building "reasoners."[19] Even in appeals to the students' minds, educators have clarified expectations about students' bodies. To support a "sound" mind, the body had to be doing something at the time, even remaining calm and still.

This chapter more fully introduces expectations for reading and expectations for students' minds, especially from Yale's math professor and president Jeremiah Day. In part, I start with Day because of the legacy of his rules of "mental discipline" for American mathematics classrooms. Furthermore, because performing math involves assumptions beyond abstracted minds, the chapter looks more closely at how math textbooks and their uses have required certain things of students' bodies: how math books surprisingly promoted early gym classes. The chapter also previews the beginnings of student movements, the seeds of frustration and disobedience within classrooms where teachers preferred extreme control. Throughout, textbooks' expectations for mental and bodily discipline (clearly expressed with regard to reading) have flagged the performative dimensions of speaking about math in American classrooms.

## Read-Throughs of the Mind

As expressed in the case of the Conic Sections Rebellion, Yale's mathematician-president Day followed strict notions of discipline, and he consistently invoked the ideals of "mental discipline" in his writing and teaching about mathematics. Such a rationale for mathematics education had its roots in Anglo-American philosophies, and it still does exert a considerable influence on education studies in Britain and the United States. The Centre for Mathematics Sciences at the University of Cambridge recently reasserted that reading about math "trains the mind."[20] British statistician Adrian Smith also noted the importance of that rationale in his "inquiry into post-14 mathematics education."[21] Though slightly less common, U.S. sources similarly follow the legacy of Day and others who claimed that math "disciplines the mind."[22] Assuming students' minds to be naturally undisciplined, these Anglo-American rationales have asserted the importance of mental fitness for professional aspirations and future civic engagement. Under Day, mental discipline showed (implicitly) how students could go out into the world and lead in government, work, and life—all through the power of reading about mathematics.

Mental discipline was already established well before Day went to school, and as he came to know, it already crystallized the intense importance of math in the education of Anglo-American boys. Mental discipline did communicate the sense of mathematics as preparatory for later life; although, in many instances, it was agnostic as to the work that the later life would contain. Rather, in mental discipline, math represented the pursuit of mental perfection through frequent practice. Just as physical exercise allowed the body to acquire superior abilities, so the argument went, the mental exercise of mathematical problems allowed the mind to improve. As Day learned, repeated mathematical exercise would confer skills that could be applied to any situation involving the presentation and interpretation of argument.[23] It did not matter that not all arguments involved math; for Day and his contemporaries, math was considered the best way to analyze any arguments and make new ones. Math provided the key to logical conversation.

In fact, the foremost mental disciplinarian, the British philosopher John Locke, had argued that the ability to transform everyday arguments into mathematical ones would not just improve a student's abilities of communication but also raise him to a position to judge other men. Locke's *Of the Conduct of Understanding*, published just two years after his death in 1704, was a handbook for becoming an autodidact: how a man could teach himself. It recommended studying algebra in order to gain the ability to weigh the assertions of others, and it suggested geometry to separate distinct ideas and lay them out in logical order. While he might not understand the whole content of the arguments presented to him, a mathematical man could at least decide which claims to accept based on structure. Of course, a certain level of consideration and analysis could occur in conversation, but it applied more clearly to instances of reading. Reading a policy, a legal document, or even a sermon, an autodidact could weigh sides, separate ideas, and understand the logic of others' words.[24] When Day and his contemporaries popularized mental discipline, they emphasized to their male students that mathematical study transformed them into leaders. Through mental discipline, reading about math was power.

Did reading about math make students feel powerful? It is difficult to tell. As later chapters explore in greater depth, records of student life are certainly incomplete.[25] Since the late nineteenth century, universities and student groups have tried to collect and publish comprehensive lists of alumni information. In ideal cases, these publications contain brief biographies of alumni accomplishments, ideally standardized from self-reports. (An excellent example is *The West Point Register of Graduates & Former Cadets*, first compiled, organized, and published by George W. Cullum in 1850.[26]) Still, even if the records are standardized—and that does not happen often—they usually just contain information about students who graduated, not, for instance, Yale's forty-five. Furthermore, what's considered important information varies a lot from one university to another. Cases of intercollegiate records, such as fraternity yearbooks, capture only a particular moment in time, such as a February in some year, not acknowledging how students enter and leave groups throughout. Finally, a small number of student records have been collected in college archives. These are usually quirky, consisting of diaries, doodles, plays, books, coffins, tickets, and programs, but they are difficult to analyze. Despite these challenges, following chapters return to student perceptions.

For now, it is enough to acknowledge the differences between the pieces of evidence and what variation in evidence can itself tell us. Students in the past can never hold still for us to understand them. It is important to recognize that not everyone graduated, not everyone joined groups, and not everyone allowed their things to be left behind. Confirming what archeologist Laurie Wilkie has said about historical studies of fraternity life, such variations are reminders for present-day scholars.[27] They remind us that studenthood was—and is—fluid. Acting like a student, reading math like a student, did not guarantee that others would perceive and acknowledge that act.

The rest of this chapter focuses on how textbooks have constructed students: who they are, and why they are studying. These statements have been important in part because they provided a framework for institutional understanding and recognition of studenthood. They do not determine the exact actions of students. They do establish rules (explicit as well as implied) for how students could and should behave. Textbooks' words have mattered in part because of the importance of reading aloud. Over generations, students have spoken certain arguments about math education. Reading the claims about mental discipline aloud, students reinforced the ideal that reading math led to better arguments.

## Popularizing Mental Discipline at Yale

A closer look at the mental discipline of math textbooks emerges through focusing on Day. His father, also named Jeremiah Day, had been educated in theology at Yale but took a roving path through the various jobs available to a well-read man in colonial society. By the time the younger Day was born in 1773, the elder had been a schoolteacher, farmer, and town representative to colonial assemblies, and he had finally settled down to a post as a Congregationalist minister. The younger Day's mother, Abigail Osborn, also valued education. The four boys born to Abigail and Jeremiah started learning Scripture, mathematics, and the classics at a young age, particularly after David Hale became the family's tutor in 1785. Hale, a Yale graduate himself, constructed a strict course of studies around the Yale entrance requirements.[28] Later in life, after his own college experience, the younger Day established and promoted justifications for math study, connected perhaps to an old-fashioned view of Yale life. Through his textbooks, he popularized mental discipline as a main reason to read about math, particularly Yale mathematics.

When Day started attending Yale in 1789, American educators thought of math as recognizably collegiate, with deep roots in European history. Since the Middle Ages, math had formed an essential part of what was called a "liberal education" at European universities. Historically, there were seven liberal studies: the mathematical *quadrivium* (four paths) of arithmetic, geometry, music, and astronomy, balancing the linguistic *trivium* (three paths) of grammar, logic, and rhetoric. Though seldom covered in equal measure, the interplay of these topics was supposed to build the basis of a university education. In liberal studies, arithmetic and geometry proved foundational for the *quadrivium*, since arithmetic built the mathematical introduction to music and geometry introduced astronomy. At medieval universities, geometry also provided a connection to ancient Greece, particularly through scholarly consideration of works of Euclid, Archimedes, (pseudo-)Aristotle, Apollonius, and others.[29] Math subtly carried these associations to Yale.

Even outside math classrooms, math encircled American colleges. For instance, it provided one way (of many) that colonial colleges could embody the dream of America's descent from ancient Greece and Rome. American colleges used medieval universities as models, and European universities in turn were looking back to ancient Greece. In college architecture, American institutions adopted modified versions of quad designs from medieval European universities, which in turn were inspired by accounts of the Athenian Academy, Lyceum, or other ancient educational sites. In curriculum, American universities and schools did not rely on national bodies to standardize offerings; there were few such organizations before the Civil War. Instead, math provided continuity with medieval forbearers and their ancient Greek heroes. Along with monuments, urban planning, and architectural development, college math fed the cultural hope that the United States was also a descendent of Greco-Roman civilization.[30]

Still, Day's student days were not so much about math as about his lingering illness. In 1791, while doing a self-directed study of theology, he became entranced with the idea of following his father into the ministry; then he had to leave school because of pulmonary trouble. The mysterious illness, which did not have an effective treatment during his lifetime, left Day searching for a career that did not excite him excessively. Returning to Yale, he tentatively embraced the modified *quadrivium*, especially geometry, and he began to devote his life to math education. Shortly after graduation, he became headmaster of the nearby Greenwich school and a mathematics tutor at Williams College, and he returned to Yale as a mathematics tutor once again. Meanwhile, Day received a license to preach, which he still gladly did despite his intermittent illness. His appointment as professor of mathematics and natural philosophy at Yale in 1801 also coincided with his most dramatic health problem: a hemorrhage supposedly caused by his overzealous theology. In later life, he became known for a cautious style of classroom management, college management, and work in math education. All these professional decisions, according to later historians, had as much to do with pedagogical interests as about his careful avoidance of health problems. Day came to math not because of intense passion for its beauty or even its theological ramifications. Instead, it appears, he devoted his life to it because it seemed detached, calm, and sure.[31]

Mental discipline, for Day, provided the perfect groundwork for a related textbook series. At the time, many educated Americans already assumed that reading about math led to logical argumentation and communication, and though borrowed from a British philosopher, views of reading provided a persuasive way to argue for/to an American audience. In the preface to his

*Algebra*, Day explained how British mathematical books were insufficient for American students. Either they were too detailed and "too voluminous for the use of the body of students in a college," or they were too basic, "*text-books*, containing only the *outlines* of subjects which are to be explained and enlarged upon, by the professor in his lecture-room, or by the private tutor in his chamber." Either way, the books required too much professorial involvement, according to Day.[32] Not only did Day borrow mental discipline from Locke, but he also constructed his textbook to be another handbook for autodidacts. Moreover, relying on contemporary stereotypes of American independence and European deference to authority, Day presented himself as updating classic math books, even ancient Greek ones, for an American audience. Claiming to have read and synthesized the works of Pythagoras and Euclid, as well as Newton and Euler, Day built on European precedents and simultaneously argued that his book was better, especially for Americans. For Day, it was akin to the American War of Independence, writ in algebraic notation.

Day ambitiously implied that reading his *Algebra* should replace Euclid's *Elements* at American colleges because of his understanding of mental discipline. His book taught the "general principles" of math to create "logic" and "sound reasoners." Unlike others, it did not begin with specific applications to business or even navigation and surveying. Instead, like Euclid's *Elements*, Day's *Algebra* began with broad statements, explored through proofs and written descriptions. It developed these through the typical framework of mental discipline. Through repeated problems, it aimed to lead students to new thoughts (new thoughts for them—not truly "original" thoughts), which would build a more complete understanding of how their "general principles" could become "distinct propositions" and then even more specific "applications." An abstracted version of the everyday, Day's *Algebra* did not apologize for its systematic formalism. Instead, it pointed to the precedent of Euclid's *Elements*, which was then revered for its similar treatment of general statements of math, carefully abstracted from the world. Both, it claimed, built a sense of the real world from "general principles," and both were therefore able to "exercise" the student's mind without "fatiguing" it.[33] Explicitly attempting to borrow the cultural clout of Euclid, implicitly adapting mental discipline from Locke, Day placed his book on par with a centuries-old masterpiece.

Taking mental discipline one step further, Day suggested his *Algebra* could also be found in all realms of human activity. He began by listing the math requirements for jobs in "*Mercantile transactions*," "*Navigation*," "*Surveying*," "*Civil Engineering*," "*Mechanics*," "*Architecture*," "*Fortifications*," "*Gunnery*," "*Optics*," "*Astronomy*," and "*Geography*." For instance, "Mathematics principles are necessary in *Mercantile transactions*," he explained, "for keeping, arranging, and settling accounts, adjusting the price of commodities, and calculating the profits of trade." Furthermore, "in *Astronomy*, for computing the distances, magnitudes, and revolutions of the heavenly bodies; and the influence of the law of gravitation, in raising the tides, disturbing the motions of the moon, causing the return of the comets, and retaining the planets in their orbits."[34] Encouraging the reader to imagine the diversity of workers doing math, Day's *Algebra* also indicated even more dramatic effects. The conflation of mathematics and gravitation, if read quickly, suggested these academic subjects could be so powerful as to control celestial bodies and keep them in their orbits.

Continuing, Day argued for the universal application of his new text, showing how reading math could be relevant to all human activities. "In *History*," he claimed, mathematics was valuable "for fixing the chronology of remarkable events, and estimating the strength of armies, the wealth of nations, the value of their revenues, and the amount of their population." And "in the concerns of *Government*, for appointing taxes, arranging schemes of finance, and regulating national expenses." Establishing the importance of math for military and economic knowledge, he resorted to listing fields without explanation: "The mathematics have also important applications to Chemistry, Mineralogy, Music, Painting, Sculpture, and indeed to a great proportion of the whole circle of arts and sciences."[35] Not only for merchants, surveyors, and astronomers, Day's *Algebra* also claimed connections to a huge variety of fields from the scientific to the social and artistic. Moreover, in describing the range of human pursuits through geometry metaphors—"a great proportion of the whole circle"—Day subtly indicated the reach of his mathematics. After all, if math could be extended to all arguments in all areas (as Locke said), it could certainly be used to justify itself.

## Reading Math, Establishing Yale

For Day, mental discipline did not support reading about just any kind of math; it supported reading about Yale math specifically. Though Day's *Algebra* was not as popular as Euclid's *Elements*, it did establish the importance of Yale math for the American textbook trade, as historian Amy Ackerberg-Hastings has argued.[36] Day's arguments about the best system for "American colleges" were clearly rooted in his own experiences as a math professor there. In 1817, the faculty and board of the college recognized Day's achievements by electing him to be the ninth president of Yale, and he soon began teaching mental and moral philosophy because of the expectation that the president would be the theological head of the college. Despite the reduction in his math teaching, Day continued to publish new editions of his textbooks, adding a presidential title to his author pages. Day's advocacy of mental discipline already justified learning math, with reading math as the first step; through his new position and through further textbooks, Day similarly built Yale's reputation.

"Adapted to the method of instruction in the American colleges" in general, Day's textbook series justified and naturalized the importance of reading about Yale mathematics. After all, its broad organization showcased Yale's updating of the medieval *quadrivium*: the four, foundational mathematical paths. Once referring to arithmetic, geometry, music, and astronomy, Yale's new "four parts" were algebra, trigonometry, mensuration (meaning measurement), and navigation/surveying. Day's *Algebra* and *Trigonometry* represented the "first part" and "second part of a course of mathematics."[37] Published in 1814 and 1815, they joined Howe & Deforest's line of books from Yale faculty, an eclectic assortment that included President Timothy Dwight's sermons and Professor Benjamin Silliman's travel narratives. Their printer, Oliver Steele, published the "third part" of the series: an updating of Euclid's *Elements* called Day's *Practical Application of the Principles of Geometry to the Mensuration of Superficies and Solids*.[38] The "fourth part," from Oliver Steele's new company Steele & Gray, reflected the new focus as well: *The Mathematical Principles of Navigation and Surveying*.[39] The course overall, from two separate publishing houses, allowed Day to argue that Yale's classes should be central to American mathematics.

These books provided other publishing houses with a new sort of venture too: a book purportedly for Americans by an American (and a Yalie at that). By 1827, Day's *Algebra* was in its fourth edition. While still published by Hezekiah Howe in New Haven, it had gained two large publishers in other cities: Collins & Hannary in New York City and John Grigg in Philadelphia. Collins & Hannary had previously established their reputation through American editions of a large number of British books: legal digests, introductions to Latin syntax, works of moral philosophy, and others. Philadelphians knew John Grigg, too, not for his American publications but rather for translations of scientific and medical texts originally published in French. Containing a note from the Office of the District of Connecticut saying that Day claimed authorship, these editions likewise indicated the popularity of his *Algebra* by noting that false copies and forgeries might be sold. Not only was the textbook an American product, but it was also a remarkably successful one that other Americans might want to steal. As many math historians have noted, Day's textbook series promoted his interpretation of Yale's educational vision and ultimately reached hundreds of thousands.[40]

Did Day succeed? In a certain sense, yes. The language of mental "exercise" from his *Algebra* has appeared so often as to be unremarkable. Would a student or educator be surprised to read or hear that algebra exercises the mind? Would she raise an eyebrow at the comment that math can be used as a tool to analyze any argument? Probably not. Moreover, Day's project of writing a textbook by an American, for Americans has become an everyday affair; few people explicitly comment on the national origin of a given textbook anymore.

Furthermore, perhaps most importantly, Yale's excellent reputation has been established, in part because of efforts to publicize curricular updates in math and beyond.

That said, there were (and are) some problems with a strict adherence to mental discipline. According to Day's interpretation, reading about math transformed students' minds. By reading math, they learned the importance of numerical skills, logical argument, and clear communication. Though the textbook series clearly assumed a changed mind, strengthened through repeated, frequent practice, they did not say much about how to communicate with other people. If mental discipline means a demand for logical communication from others and an adherence to logical communication from oneself, what of the people in the exchange? How should they speak? How should they look at each other? What should their bodies be doing? Through similar questions, another New Haven teacher a generation younger than Day began to expand the growing educational focus on mental discipline and Yale mathematics. Where Day focused on the mind, his successor focused on the body.

## Math's Bodily Discipline in New Haven Schools

Concentrating on the case of young white men, American educators of the 1820s–1840s increasingly asserted that students would be happier and more productive if given a new academic requirement: gym class. Reading, writing, and arithmetic necessitated a relatively stationary posture, and they therefore had the reputation for making students weaklings. As historians have observed, the gym movement asserted how physical exercise, instituted at the school or college, would not only counteract the effect of studying but also promote healthy exposure to fresh air and muscular work.[41] Not to be confused with manual labor, as educators argued at the time, gym attempted to replicate the performances of gymnasts, using the props of rings, bars, supports, and bags. Its curriculum likewise extended to intramural sports and even military drills, not because students were expected to become professional athletes or soldiers, but because they were expected to follow rules. Gym, in sum, assumed performances of deference and obedience, which could then be applied in academic learning, particularly for mathematics.

In certain educational contexts, math's mental discipline became tied up with theories of physical exercise and required gymnastics. According to the popular terminology of the day, the "exercises" of proofs and math problems strengthened the mind just as routines with uneven bars strengthened the body. It was commonly believed that the mind, theoretically partitioned into its own muscles or "faculties," needed its own academic gymnastics, and educators regularly claimed that math was the best subject to provide them.[42] Math featured linguistic patterns, repetitive steps, argumentative structures, and difficult puzzles, all of which caused students' minds to "sweat." Likewise, it worked all their mental muscles and relatively equally too. Math promoted the goals of symmetry and balance for the mind, as gym exercises did for the body. In fact, gym provided motivation for scholastic work precisely because of the assumption that students would seek proper proportion. Once their bodies were strengthened, their minds would pursue strength as well. Bodily discipline inspired mental discipline and vice versa.

Another New Haven teacher a generation younger than Day, Yale alumnus Nathaniel William Taylor Root, particularly connected mental discipline and required gym through Yale mathematics. In 1857, Root published a book about math, exercise, motivation, and student discipline called *School Amusements; or, How to Make the School Interesting*. At the time known for his inspired teaching of math, Root soon became a figure of the gym movement: his *School Amusements* spent only a few pages on mathematical games and otherwise argued for the importance of physical exercise through military drills and required gymnastics. Root's book therefore exemplified the combination of math, exercise, and discipline, advocating math's mental discipline through military drills.

In 1857, N. W. T. Root was an unlikely spokesman for innovative teaching. Just twenty-eight years old, he was more of a perpetual student than a natural teacher. Growing up in New Haven, where his father was a minister, Root entered Yale only for a year, when he was nineteen. He completed the freshmen studies, and he promised to continue training for the clergy but not there. Taking occasional classes in theology throughout his twenties, he meanwhile supported his small family by being a schoolteacher. Though not his chosen profession, teaching offered him various opportunities in New Haven and surrounding towns in Connecticut. Root attended pedagogy workshops, went to educational conferences, and read the latest books on the subject.[43] Applying his theological devotion to education, he developed a good local reputation, particularly for his treatment of the usual school subjects that relied on rules and memorization: arithmetic and geography. He worried extensively about motivation too—making school more interesting for students and teachers alike. In fact, he began to publish on the subject. Though few people knew it at the time, he had also published once before *School Amusements*: an anonymous volume about Yale college life. Though a full consideration of Root's secret book will need to wait until chapter 2, that book also took math as key to addressing the "dull routine" of school and college.[44] His public book connected math to physical and military exercises: using assumptions of mental discipline, it suggested math drills.

Mental discipline's assumptions motivated the "amusements" in Root's *School Amusements*. The only scholastic activities labeled as "games" appeared in the last 8 pages of the 225-page book. The first, about geography, came from a meeting of Horace Mann's Teachers' Institute, which Root attended around the time he started college. Mann asked the 150 participants, both "ladies and gentlemen," to stand in a circle, placing their backs to the walls of the room, and he explained that the game was about cities and states. Starting with "Boston, Mass.," he asked the nearest person to come up with the name of a town that began with the final letter of Boston ($N$). That person, saying "Newport, Rhode Island," then prompted the next participant to find a city beginning with the letter $T$. Mann told the group that no one could repeat the name of a town previously given. The group did compete successfully, but as Root remembered, "the last dozen of the hundred and fifty were somewhat puzzled to think of new towns, and others were in momentary difficulty when the letter which came to them was a $Y$ or a $K$, or a $Z$. Towns which begin with $E$ also became scarce."[45] The little "game" assumed that participants would be better students because of repetition and quick thinking. Though about geography, it did borrow from mental discipline's ideals about building better students. Foremost, it implied a stronger mind would follow from mental exercises done again and again.

Competition provided a missing component for Root: games were fun or "interesting," as he put it, because everyone tried to be the winner. He explained the idea by discussing the second round of the geography game. It added a time limit of one minute; those who took too long for their response had to sit down. These rules brought the group of 150 down to about seventy-five. Then after another go around, only ten remained, "six ladies and four gentlemen." At a certain point, Mann increased the time limit to two minutes, but he did not lift the other restrictions. After many minutes, a few had to sit down, and then only two remained: a female teacher and a male one. The battle of "sex against sex" ended with the man, assigned a $K$, jokingly suggesting "Kalcutty" and in essence forfeiting.[46] Such an "exercise," concluded Root, could work in any school setting, not only about towns but also about rivers, mountains, atlas descriptions, or even adjectives and events. For Root, unlike for Day, it was not enough for students to be reading, even aloud. Without specific rules for the involvement of other people, there was not much point in building a better mind. Specifically, competition heightened engagement; competition encouraged students' mental discipline.

Root used the geography game to suggest math drills, now a staple of American classrooms. Similarly expanding on the usual ideas of mental discipline, he suggested how small battles could motivate students to be better prepared (mentally) for arithmetic. He remembered how, when he was a boy, his teacher would stop class before the end of their day and ask them to take out their slates. Rapidly proposing complicated calculations, the teacher would reward the first boy who answered: he could leave school first.[47] Presumably, the "exercise" continued until only the teacher remained or the school day ended, whichever came first. Root applauded the game's goal of providing "practice in rapid calculation." After all, the repeated action of "rapid calculation" caused students' minds to sweat, making them better at math through repetition and exertion. Competition provided new rules about how math involved interactions among people.

Even in the middle of advocating math drills, though, he also did begin to propose some problems with implementation. He worried that it would not be fair in "almost all schools" where students had varying mathematical abilities.[48] There, he said, students could be ranked into two or three groups, then given calculations of varying difficulty. Reaching a possible answer, a student would have to raise his hand, and if his verbal description and written explanation were correct, he could go. Anticipating logistical problems with the arrangement, Root did not give a specific plan. How would a teacher control two or three math drills at the same time? How would he (the teacher, for Root, was always "he") listen to two or three answers at the same time? How would he keep proposing appropriate problems for the two or three groups at appropriate times? What would allow the students to pay attention to the right level for their group? Perhaps he did not continue because he discovered a fundamental problem with his educational philosophy. Mental discipline assumed all students would improve in similar ways; it was not careful to point out variations. Because it was about autodidacts, it certainly did not assume roomfuls of students who could be different in their reading and comprehension.

Backing off from these concerns, Root ended with general justifications for academic competitions. Since everyone assumed the importance of mental discipline, it was not necessary to go back to ideas of repetition and exertion. Instead, he brought it all back to motivation, back to the idea that he was proposing: "how to make school interesting." As he said about the geography activity, "The teacher who adopts [this exercise] will be pleased to notice the eagerness with which his scholars, after once learning the *modus operandi*, will examine their maps, to prepare for the next trial."[49] Such an academic drill did not just inspire performance at the time; it inspired preparation, reading (of maps, for instance) in order to be the best in the class. Also, in the case of math, students had to study to do the calculations and get the reward. Being victorious meant being dismissed because the winning students were supposed to be the ones who would keep studying and keep reading, even beyond the bounds of the drill.

Math drills disciplined, in the early formulation. They exercised students' minds. They provided the occasion for repeated action, sort of like the mind lifting weights. Students, not actually consulted about their views, were expected to find such a "game" fun or at least were supposed to go along with it. Beyond mental discipline, math drills provided extra rules to govern their class time, apparently inspiring continued study—that is, further reading of mathematics. Math drills also communicated certain expectations for obedience, ones that sounded almost military.

In fact, math drills did come from military drills in *School Amusements*. Beginning with the motto "Every Teacher His Own Drill-Master," the book contained nearly one hundred pages about military drills and another fifty about "gymnastics": suggestions about school sports as well as routines with parallel bars, swings, rings, boards, ladders, pegs, horses, and ropes.[50] Physical exercise, Root claimed, would instill proper discipline in groups of boys, and they would want to do it too. He suggested that even before approaching school boards or overseers, the teacher should approach his students about the idea of mandatory military drills. They, he explained, would get excited about acting like soldiers, having uniforms and (toy) weapons, and preparing a public parade at the end. A continuation of academic competitions—an analog of math drills—these military exercises would supposedly strengthen the student body.

Root's publication strategy further emphasized the connections between math and military exercises. A teacher known for his fascination with geometry, Root published his book with A. S. Barnes and Company, a publishing house that had its origins in the math textbook trade. With its main office in Connecticut, it had grown from a local printer to a national player through selling the series of school and college math from Charles Davies, West Point teacher extraordinaire. Though the company had expanded to general books about teaching by the time Root approached it with his manuscript, A. S. Barnes was still best known for its mathematics, including *Davies' Practical Mathematics*, which Root highly recommended.[51] Root's audience was made up of the teachers who found arithmetic "dull . . . dull," and he offered how students could get interested in math through their fascination with West Point and especially soldiers.

These school activities, in short, mixed math with military-style obedience. With permission from parents (or perhaps without), the teacher could write his own "pledge of allegiance," the students' "promise to obey" the teacher. With properly obedient pupils, the teacher could then introduce twenty drills that could take the new "squad" from standing silently ("fall in") to marching in arches, circles, ovals, and winding paths. The end product, for Root, would combine the spectacle of school discipline and physical fitness with the embodiment of geometric forms. Literally circling the teacher, the "companies" advertised the value of the school by showing how well they could obey rules, both military and mathematical.[52]

## Performing Math Class and Gym Class

Where the military drills communicated obedience to teacher and geometry, the public performance of gymnastics showed strength: physical exercise as well as the mental exercises of math drills. "It is a fact not sufficiently noticed and lamented, that young men of this country," began N. W. T. Root, "are, as a class, but weak and effeminate specimens of manhood. We see them on their way to or from the counting-room, the office, the study, dragging along their half-vital frames, pale-faced, dyspeptic, sacrificing themselves to gain a fortune which they may not have life . . . to enjoy if obtained."[53] Saying that he heard of schoolboys in England who were as fit as sailors, he asked, where could such "specimens" be in New England? Continuing in the mode of pseudomedical pontificating, Root suggested that gym class could also be a public exhibition of the boys: showing them to be natural, beautiful, and powerful. He therefore recommended the students be encouraged to doff their shirts and tighten their belts while in gym class outside. Such activities, even more than drills, would demonstrate to the public how teachers could "make them *men*," providing "a thorough education, physically as well as mentally."[54] The boys, tired from their runs and seeking symmetry in their development, would leave the field ready for mathematics. Throughout, gym class justified math class.

Root argued American students needed gym because he thought it would strengthen the reputation of math teachers. He criticized not only the physical "specimens" of American students but also the specimens of American teachers. Some teachers, he observed, found teaching too "dull" to interest them, hoping to stay in it just long enough to establish another career. Others complained of the "dull routine," not knowing about "modern improvements" in pedagogy. Yet others were young, untried, and desirous of advice. He claimed he wrote *School Amusements* to satisfy all three audiences, showing how their students could be walking advertisements for their schools, raising the status of teaching and raising teachers' pay at the same time. The trick, for Root, was not to avoid difficult subjects but to make schoolwork, "*hard*" labor," "*seem play*."[55] Games and exercises lent a sense of fun and adventure, from gym class to math drills and back again.

The student, for Root, was an ideal body inscribed in geometrical shapes; Root's vision of physical exertion was literally surrounded by mathematics.[56] Written for math teachers, by a distinguished math teacher, *School Amusements* began with complaints about teaching's "dull" routine, and it ended with rules for an arithmetic competition. Even the numerous descriptions of sports, stretches, drills, and the like emphasized the performance of fitness and health, showing students' bodily forms through their ability to replicate geometrical shapes. Circling the teacher or the "sergeants" or "captains" who stood in for his authority, the class showed themselves to be obedient and powerful. Even the gym student performed mathematics.

Students needed to seek proportion not only in their bodies but also in their minds. Like gym, math let students compete with each other, leading to mental as well as physical development. Alongside gym, arithmetic became another venue that rewarded competition, speed, and ability. The word "exercise" was at the heart of *School Amusements*. It connoted duty and worship, military drill and scholastic training, physical exertion, and public exhibition. In both gym class and math drills, exercises made sure students developed well, both physically and mentally.

Root's view of math education confirmed the emerging expectations for performing white American masculinity. As historian Harvey Green and archeologist Laurie Wilkie have observed, fitness movements and educational cultures of the time changed to match the changing notions of what it meant to be perceived as a young, healthy man.[57] Similarly, in various subtle and obvious ways, Root assumed all the teachers he addressed and all the students to be seen as performing appropriate (white) manhood. Though their faces were not supposed to be too light, because that would mean too much time spent indoors, Root still did assume their natural coloring to be pale, and when he compared them with English students and teachers, he expected the Americans to be similar to British noblemen, lawyers, clerks, and shopkeepers. In other words, they were from the upper and middle classes and, given the comparison, likely of English ancestry. Moreover, despite the anecdote about "lady" teachers at Horace Mann's geography workshop, Root used male pronouns throughout when he addressed teachers and students. The fundamental motto of the book, "Every Teacher *His* Own Drill-Master" (emphasis added), served as a call to men in particular. After all, the math teacher had to show himself to be manly, garnering respect and obedience from groups of (sometimes shirtless) youth.

Through such connections to gendered performance, *School Amusements* was an early meditation about teaching math to boys. The military drills, according to Root, needed a certain degree of "brotherhood" to be pulled off successfully. The gymnastics routines needed to be led by a young man whose actions the "boys" would want to emulate. Even the arithmetic game emphasized a competitive spirit, which Root asserted to be particularly "natural" in boys. Variously equating leadership, emulation, and competition with men, Root built a system of boys' education inspired by math and executed through physical exercise. As in more recent books about teaching math to boys, *School Amusements* equated students' displays of success with their conforming to norms of manliness and healthy physical development, as well as their victories in mental games and competitions that rewarded quick calculation.[58] Root's student ideal linked gym class, math drills, and performative American masculinities.

Surprisingly, given the topic, Root kept a number of his exemplar "amusements" secret: he nowhere mentioned his pedagogical game of writing silly verse about mathematical terminology, a fundamental feature of his anonymous publication. *School Amusements*, after all, assumed it "natural" for boys to want to emulate the behaviors of manly men. Teachers took on the guise of soldiers, athletes, and gymnasts and prompted student obedience through the boys' desires to do the same. Within the setting, singing and writing poetry were not exactly womanly, at least not in the "unhealthy" way that Root periodically worried over. However, these activities did not have the same manly status as gymnastics and military life. Some instances of student fun stayed secret, even in Root's publications.

## Reading Math from Gym

Though gym (especially Root's version of gym) did not involve much reading, gym class still did promote reading mathematics. According to the mental discipline thesis, repeated, rapid drilling on math problems would cause students' minds to sweat. Assuming that students would want symmetry/proportion in all things, the mental work would encourage them to do physical exercise.[59] After marching with classmates, making various shapes on the field, they would be tired and want to learn back in the classroom. Exercising outside would encourage them to do exercises from math textbooks inside. Math class led to gym class, which in turn led to math class. Gym class, in short, not only promoted reading math; it also came from mathematics.

In these ways, reading math did involve embodying math through performance. The ideal student (fit and proportionate in his body) studied math in the classroom and then went outside and marched (shirtless) in various geometrical forms. He was surrounded by math, literally inscribed by its lessons. The best gym teacher, for Root, was a math teacher because he presumably provided these reminders, if not verbally, at least through his example. The teacher's symmetry/proportion in both mind and body allowed him to be the "drill-master" in military exercises, math drills, and gym class alike. He embodied geometrical proportion in his literal form and in the communication of mathematical ideas.

Math communication then has involved assumptions about students' bodies. According to the ideal of studenthood written into these textbooks, it was not so much about presenting orally or speaking at an appropriate volume. Neither was it much about when to start speaking or how to keep going. Similarly, there were few instances of conversation among students, particularly beyond the bounds of scholastic and athletic drills. Instead, in these idealized textbook exchanges, teachers gave "orders" to students, and students were expected to participate obediently.

Instances of disobedience, of course, did occur. Chapter 3 returns to Yale's Conic Sections Rebellion, and Root's anonymous book will be in chapter 2. In fact, regarding performative math communication, the whole dichotomy between obedience and disobedience has been a little blurry. Cases of clear disobedience (even burning books) can be signs that students are understanding and rehearsing their knowledge. As I explain in chapter 2, thousands of students did reuse information from their math classes in many institutionally discouraged events.

Chapter 1 has clarified how textbooks have made a large number of assumptions about student behavior. Students were expected to be disciplined, obedient, exercised, orderly, and developing. They were expected to strive for some ideal in their minds and bodies. Fundamentally, they were supposed to want to be like their teacher or professor. Of course, these were the ideals constructed in books written by teachers and professors. Such views said little about the actual experiences of students and the ways that they received these messages. Even beyond the cases of Day and N. W. T. Root, some students did go on to be math teachers themselves. Perhaps they did want to be like their teachers, perhaps repeating some of the same stories they heard before. Still, as in chapter 2, many students chose to resist and act differently, rehearsing some of the same messages but in increasingly prohibited ways.

## 2

# How Math Communication Has Been Practiced in Prohibited Ways

• • • • • • • • • • • • • • • • • • • • • • •

Actual students have controlled how they participate in learning mathematics. Rebellions and other instances of clear disobedience have happened over the years, though historians and education scholars rarely analyze these events within broader systems.[1] As I argue, there is a symbiotic relationship between formal classes and what teachers have called distractions or pranks. Math classes' calls for obedience, as in chapter 1, have existed hand in hand with prohibited student activities. In fact, the math classroom is not the only place where actual students have practiced math communication; in making up silly songs, in doodling, and even in destroying school property, they have rehearsed what they learned from reading math textbooks.

In fact, some instances of student disobedience are analogous to theatrical rehearsals for math communication. In theater, rehearsals happen after read-throughs, and they last until the first night of the performance. They gradually build components into a show: how actors should move around a space, how they should embody their characters, and how they should start to use costumes and props. In the rehearsals for learning math, students learn the same sorts of things: the movement, embodiment, and even dress and props appropriate to math communication. As noted in chapter 3, chalk dust (and being covered in it) has been a key component of math education for generations. Rehearsals for learning math have incorporated similar props, though many have occurred beyond the classroom.

This chapter participates in the growing scholarly recognition of "student cultures." In one notable example, historian Helen Lefkowitz Horowitz created a rough taxonomy of undergraduate life using memoirs, letters, fiction, and other histories.[2] Claiming that the groups had not changed much in the two hundred years from the 1780s until her book came out in 1987, she called out the college men, outsiders, and rebels. Youthful and immature, the college men opposed faculty calls for obedience, ultimately facing the most dramatic change when the Civil War and its aftermath forced coeducation on their masculine enclaves. The outsiders did not go along with the college men. They did not participate in campus rebellions, pranks during chapel, or other movements against military drill or against prayer. Pious and serious, outsiders became the "dull" students and then "nerds" of the twentieth century. Finally, the rebels combined elements of each. Though generally respecting their professors, they nevertheless aimed to win at college life, aimed to be college men. Yet something held them back. Because of their religious background, their ethnic appearance, their socioeconomic origins, their politics, or their gender, they could not join Greek life and/or they could not participate in core college events. Their stories have provided the greatest continuity, telling us about segregation and exclusion in higher education, how institutional biases have changed over time.[3] Taken together, these groups inspire a framework for viewing actual student experiences of math communication at both the school and college levels. Generations of students (especially college men) gained status through elaborate math pranks.

Just as chapter 1 argued for the ways learning math has involved the body as well as the mind, chapter 2 is about how math communication has been practiced beyond the classroom. Such a project must also expand beyond historical analyses of traditional schoolwork. Various other routines and rituals have marked the academic year for generations of students—convocations, examinations, concerts, sports games, religious services, newspaper/yearbook/magazine publications, commencements/graduations, and campus traditions. What Lefkowitz calls "college men" have tried to place themselves at the center, especially of the campus traditions, making it seem they control the whole school. As we will see, though, they were not entirely successful at pushing aside the math classroom. Even in Burials of Euclid or other increasingly prohibited activities, students practiced the knowledge and techniques they learned from math textbooks and teachers.

## Student Cultures of Math Communication

The focus on math communication and the experiences of actual students recovers some forgotten campus traditions: informal performances that focused on mathematics. Though it may seem surprising today, students at many nineteenth-century colleges held elaborate "funerals" for their math textbooks.[4] They burned their books. They consumed their books in flame to mark their metaphorical consumption of knowledge. These occasions celebrated accomplishment and commemorated class identities, and they ostensibly resulted in the destruction of print material. Yet they also generated funeral paraphernalia: tiny, textbook-sized coffins, winding sheets, pyres, urns, and even printed programs. Paradoxically, an event about irreverent destruction generated college mementos by the thousands. College libraries at Yale, Hamilton, Bowdoin, and Bates have saved such documents related to students' experiences of learning mathematics. Beyond artifacts of the classroom and the textbook, some stories of learning math have been etched in half-burned textbooks and morbid programs.

Funeral programs for math textbooks, even gathered together from multiple schools and colleges, are difficult to interpret. They used the terminology of advanced math and also a smattering of religious references. After all, students constructed them to seem like actual funeral programs, except written in mathematical allusions. They contained commemorative titles. They indicated the community of mourners. They listed participants, especially the student preachers who delivered eulogies and elegies. They diagramed the proper order of people participating in the funeral procession. They indicated directions to the gravesite. Some included quotes to inspire musings on death and memory. All devoted much of their text to hymns.

Yet the whole funeral was a farce surrounding the central math textbook. For instance, the hymns and speeches were rewritten to joke about the terminology of analytical geometry. Bowdoin's class of 1854 provided a particularly striking example when they rewrote "Auld Lang Syne" with a prominent list of mathematical terminology: "His *functions* true *transcend* all praise. / His *forms* fill every *space*. / His *curves*, in *spiral* wreaths, / Are elements of grace." Singing not only these terms but also "*gravity . . . time . . . constancy . . . Volume . . . winding curve . . . rolled . . . spiral . . . point . . . parabolic*," these juniors chose common words with specialized meanings, which they indicated through italicized print.[5] Bringing together funeral practices with math allusions, these textbook burials combined students' understanding of religious observance and math communication.

Analysis of these documents therefore participates in the scholarship that links technical communication and religious expression. Though written in the language of math, students' pamphlets consistently referred back to actual funeral documents and actual understandings of death and imaginings of the afterlife. Religion therefore provided a main lens of analysis or rather "cultural variable," as technical communicators Nancy Hoft, Elizabeth Tebeaux, and Linda Driskill have indicated in other cases.[6] As Hoft, Tebeaux, and Driskill implicitly have argued, technical communication also does not necessarily explain religious experience; stories of the divine often fall outside the historical record. Mathematical funeral programs, regarded as a parody of religion, bring in even more disorderly associations. These programs instead indicate the complexities of student culture, especially surrounding math communication.

Consider "Drill Master" N. W. T. Root's anonymous publication. As explained in chapter 1, Root publicly published a book about math, exercise, motivation, and student discipline called *School Amusements; or, How to Make the School Interesting.* At the time known for his inspired teaching of math, Root soon became a figure of the gym movement: his *School Amusements* spent only a few pages on math games and otherwise argued for the importance of physical exercise through military drills and required gymnastics. Root secretly published an earlier book too. With classmate J. K. Lombard, he compiled snatches of what they called "college poetry," much of it about geometry. Their *Songs of Yale*, what historian James Lloyd Winstead calls the "first known American college songbook," featured the strange rituals of Yale students, particularly their surprising ways of interacting with math classes.[7] Ultimately, Root not only encouraged engagement with math through "exercise"; he also shaped fundamental expectations about what it meant to take on the role of the math student.

Song-smithing became a common way for nineteenth-century college students to practice their math communication. Root and Lombard's "Burial of Euclid" songs inspired thousands of college students not only to compose but also to burn and bury their books. Such student-centered events indicated their change in status and offered an opportunity to speak back to activities from the math classroom. Variously called the Burial of Euclid (at Yale), the Burial of Calculus (at Hamilton), the Burial of Mathematics (at Bowdoin), and the Burial of Anna Lytics (at Bates), the tradition communicated a sense of accomplishment through featuring parodies of the terminology, methods, and assumptions of math classes. It marked the transformation of the target group into "college men" by presenting performances of distinctively collegiate knowledge in the streets of the surrounding towns and before audiences of underclassmen. Burning math books has built American student experiences.[8] Actual students have practiced math communication far beyond the textbook and the classroom—even unexpectedly in their destruction of school property.

## Songs of Math in American College Life

Singing, according to historian James Lloyd Winstead, has been a major force in constituting student identity from the first American colleges to the present day. Breaking the chronological development into eight parts, Winstead has argued that college singing participated in—and constructed—educational movements, from Puritanism and other traditions of sacred worship (in the seventeenth and eighteenth centuries) to the rise of student organizations and student books (in the nineteenth century) to the demographic expansions of college admissions, the professionalization of music, and the growth of intercollegiate sports (in the twentieth century).[9] In doing so, Winstead's analyses have exemplified some new ways of studying student cultures, according to his colleagues.[10] Moreover, nineteenth-century American college songs were peppered with equations and other math terminology. Through song, students practiced what they learned in math class, creating a show of American college life.

The earliest references to math singing were at Yale. Even before N. W. T. Root taught in New Haven, the Burial of Euclid gained a reputation as a feature of Yale life that combined geometry, singing, and resisting authority. The anonymous *Sketches of Yale* from 1843 included a section about it in "Part III College Life" after accounts of the history and present state of the college, heavily from the perspective of the college and class societies. "College customs," such as the Burial of Euclid, corresponded to "landmarks" in the memory of college days. Like freshmen hazing, elections to college societies, unofficial parades through town, and town-gown fighting, the Burial of Euclid was supposed to be a secret from faculty. Not only was it an opportunity for math jokes, but it also contained speeches, parades, opportunities for textbook burial or burning, and (temporarily marked) graves. Fundamentally, according to *Sketches of Yale*, it provided an excuse for secret singing. But in 1843, following the rebellions of the 1830s, Day and the faculty passed an ordinance against it, likely viewing it as another "combination." In mock-epic language, the author of *Sketches of Yale* bemoaned that the "romance of college life is gone." Indicating the complexity of the occasion, he joked that the new laws were apparently "against the burning, burying, or other celebrating the end of Euclid."[11] In other words, the Burial of Euclid was not just an attempt to resist college authority and not just an annoyance to New Haven residents: it was a student celebration of math communication.

By the end of the nineteenth century, the Burial of Euclid came to represent all student "customs" whether at Yale or elsewhere. Not only did the Burial of Euclid appear in early accounts of student life at Yale; it also came to represent "college customs" generally, as in the guidebook *A Collection of College Words and Customs*. Accepting the earlier claim that the Burial of Euclid was the "oldest distinctively student custom," *College Words and Customs* also noted the tradition's propagation throughout American academia. By the 1850s, students at Hamilton College observed the Burning of Convivium for their ancient Greek text, and those at the University of New York (now New York University) burned Zumpt's *Latin Grammar*. Still, as *College Words and Customs* said, math remained central to student celebrations at many colleges, and some borrowed the Yale tradition wholesale. At Williams College, for instance, students met in their math class's recitation room, joking about the dead "body" of Euclid then bringing an effigy to the woods for burial. Still acknowledging Yale's example before them, they considered their Burial of Euclid central to their college's identity: after all, Williams's observance had been a feature of the book *Sketches of Williams*, a clear counterpart of *Sketches of Yale*.[12] Through song, too, the Burial of Euclid came to be considered a student custom claimed not just for Yale but for American college culture as a whole. Burials of Euclid helped forge the characters of American college life through nocturnal singing about mathematics.

## Presenting Euclid at Yale

Math singing kept returning to examples from Yale. Yale's math classes, as we saw in chapter 1, provided powerful visions of ideal students, grounded in textbooks and teachers' institutes, though not necessarily in actual student experiences. A curricular emphasis on discipline/obedience led to the establishment of math drills and model textbooks, though also some feelings of unrest and frustration. N. W. T. Root captured both of these aspects. Despite a public motto of "Every Teacher His Own Drill-Master," Root also documented Yale's college traditions, especially the prohibited, math-oriented ones. From Root and far beyond, these math songs served as a way to practice the character of student life.

In Root's anonymous publication about Yale life, math was literally at the center. Published four years before *School Amusements*, *Songs of Yale* represented the work of two anonymous "compilers" of "cherished customs," later identified as Root and classmate Lombard. *Songs of Yale* emphasized the ways that Yale students regularly wrote their own lyrics to common songs, and it used samples from the classes of 1848–1855 to tell the story of students' experiences at Yale from initial visit to final "parting." The eight sections of what Root and Lombard called "college poetry" proceeded roughly chronologically through a student' college journey: after an introductory section, there were biennial songs, boat songs, Burial of Euclid songs, football songs, songs of the spoon, presentation songs, and parting songs. Assuming a possible audience beyond Yale men, they clarified nearly all the college terms in a glossary of "explanations." For instance, the *wooden spoon* was the mock trophy given to the biggest glutton in the dining hall and/or to the junior with lowest marks, and its presentation came with raucous, ceremonial singing. The Burial of Euclid, a distinctly math tradition, appeared literally at the center of *Songs of Yale*. An "old" tradition, it consisted of secret sophomore rituals, public parades, fire, and the burial of "Euclid's" ashes.[13] Unlike the traditions surrounding football, boating, or the wooden spoon, it showed how students could act creatively in joking about classroom learning, specifically about mathematics.

There was also a high level of secrecy surrounding the Burial of Euclid because of its sordid history. Root and Lombard, though not revealing their roles as "compilers," did name most of the authors of other songs. In their preface, they noted that four Yale men provided five original verses for their book, and simple cross-referencing revealed their names to be W. W. Crapo and F. M. Finch, as well as Root and Lombard themselves. *Songs of Yale* named these four, as well as numerous classmates, in nearly all sections, but the Burial of Euclid songs remained entirely anonymous. The Burial of Euclid had only been described once before: in an 1843 account of Yale student life that claimed the faculty had banned all future ceremonies. Perhaps unsurprisingly, *Songs of Yale* revealed that the students did not listen: the Burial of Euclid continued for the classes of 1853, 1854, and 1855. Merely listing the authors of the songs as "Anonymous, '53," "Quivis, '54," and "HowAreYouNow, '55," the songs did not reveal exactly who flouted college authority.[14]

Despite their anonymity, the Burial of Euclid songs were central because they captured, for Root and Lombard, what it was like to be an actual Yale student. Actual students sang and wrote "poetry," even when it was against the rules. Actual students disobeyed their professors and tried not to be caught. Actual students respected tradition but not strict discipline. In microcosm, the Burial of Euclid songs served as Root and Lombard's impressions of the characters of Yale life. In creating new lyrics for old tunes, they demonstrated what they thought of as popular songs at the time: a medieval students' drinking anthem ("Gaudeamus igitur"), a tune made famous through blackface minstrelsy ("O Susanna"), and a hymn possibly connected to the German singing school movement ("O, pueri me longum ferte").[15] Scholastic, theatrical, and religious all at once, the Burial of Euclid songs intersected with many aspects of college life, creating a complex educational vision centered on math communication.

Most of all, like the biennial examination and presentation day, Burials of Euclid allowed actual students an opportunity to show off. The oral examinations of sophomores and seniors (called "biennial examinations") drew crowds of spectators from the town and gown alike. Similarly, presentation day introduced the seniors who passed their second biennial to the college president, faculty, and guests as the men qualified for AB degrees. Such traditions explicitly encouraged sophomores and seniors to exhibit how much they knew. By doing so, they joined the communities of Yale students and alumni, and they celebrated their achievements by celebrating their college.

The Burial of Euclid songs, though containing lines about their hatred for their college, their instructors, and their studies, allowed sophomores to show off their math knowledge too. Moreover, unlike the biennial examination and presentation day, it did not descend from educational authorities. In the Burial of Euclid, sophomores simultaneously destroyed school property and kindled school spirit. They presented themselves in the New Haven streets as college students, and they meanwhile made fun of such public presentations, at least the ones that were officially allowed.

For these math students, what was college? At first glance, it seemed a series of inside jokes—mainly drawn from the textbook and the classroom. The parody of "O, pueri me longum ferte" imagined sophomores traipsing through a winter morning and prompting each other to "elucidate" *reductio ad absurdum*, a common proof form in Euclid's *Elements*. Similarly, it described their final destination of drinking beer together while talking about "theorems hard and dry."[16] The last verse of "O Susanna" went even further: combining references to new classroom technologies (the blackboard), Euclid's British translator (John Playfair), Yale's teaching staff (tutors), and simple geometry terms (points, square).

No more we gaze upon that board
Where oft our knowledge failed,
As we its mystic lines ignored,
On cruel *points* impaled.
We're free! Hurrah! from Euclid free!
Farewell, misnamed Playfair,
Farewell, thou worthy Tutor B.,
Shake hands, and call it *square*.[17]

According to these songs, learning math at Yale was a matter of reading proofs and learning new vocabulary. It meant coming together and practicing their math in classrooms and beyond.

More specifically, the Burial of Euclid songs emphasized how individual students became a collective "we" through learning and then mocking the Books of Euclid. Entirely in Latin, their parody of "Gaudeamus igitur" took the tune for a medieval drinking song and made a story about the "old mathematician," Euclid. It began

Fundite nunc lacrymas
Plorate Yalenses,—
[Fundite nunc lacrymas
Plorate Yalenses,—]
Euclid rapuerunt fata
Membra et ejus inhumata.
Linquimus tres menses.
[Linquimus tres menses.][18]

Roughly meaning "Scatter your tears / Cry, you Yale men,—[Scatter your tears / Cry, you Yale men,—] / Death has carried off Euclid, / And he unburied / We quit in three months [We quit in three months]," it implored listeners to join in a mock funeral for geometry. About "Yalenses" ("Yale men"), their song proceeded to invoke the seniors, juniors, freshmen, and faculty in later verses, inviting the whole college body to participate in their celebration. Celebrating math, for them, made class and college solidarity possible.

Burial of Euclid songs were at the heart of *Songs of Yale* not just because they demonstrated the draw of prohibited (math) celebrations; they also showed the clear downside of college life. From a student perspective, college life was so fleeting. The summer beers, winter hikes, spring examinations, and autumnal geometry classes would come to an end. In fact, Root and Lombard hoped to use *Songs of Yale* to hold on to some of the ephemeral stuff of Yale experiences. "Almost every student has a loose collection of Songs which he has laid up from time to time, and which he values as mementoes which will hereafter bring his Alma Mater and all her cherished customs vividly to mind," they wrote. "But detached papers are not a very secure possession."[19] In order to provide greater security and solidity to their and classmates' ephemera, Root and Lombard claimed, they decided to publish an anonymous book. If nothing else, they joked, the pages would be bound together, unlike the loose pages of their classmates' recollections. Therefore, the Burial of Euclid epitomized the problem of fleeting college years. It both provided a route to class solidarity and ended in ashes of burned text. In publishing these songs, Root and Lombard tried to make sure that the bonds of class friendship did not, like the textbook's residue, blow away or go to ground.

In *Songs of Yale*, as in *School Amusements*, math served as the point where various aspects of school/college life converged. At the school level, math drills fundamentally showed how studying could be interesting for teachers and students alike. Through *School Amusements*, Root connected arithmetic not only to other academic subjects but also to physical exercise, military drill, student obedience, instructional leadership, and ideals of manly behaviors and bodies. At the college level, math's traditions epitomized the reasons for a college songbook. There, Euclid intersected with math terminology, prohibited ceremonies, class cohesion, and popular singing. In short, Root's vision of learning math made possible the rest of school and college life.

But how idiosyncratic were these views? After all, N. W. T. Root was only one math teacher, and he did not even stay in teaching for his entire career. As he promised, he did finish divinity school. In fact, he did so soon after publishing *School Amusement*, and he went on to a string of appointments as a preacher, chaplain, and rector in Rhode Island, Connecticut, Maine, and eventually the Union army.[20] Lombard, Root's partner in poetic anonymity, followed a similar path, completing his studies at Yale. After acting as a teacher and then principal at various schools in New England and New York, Lombard too pursued his ordination and became a rector in Connecticut and Massachusetts.[21] These two would-be preachers helped popularize math singing and spread the good word about burning books. For them (and the thousands who followed their example), learning math was central to student life. Textbook recitations and other classroom experiences did matter, mainly because they offered a shared experience. On the same order as winter hikes and summer beers, math classes forged the characters of student life: ones that could be shown off in funerals for math books at Yale and elsewhere.

## Impersonating Calculus at Hamilton

The ways alumni of other colleges told similar stories can explain further how math communication has been practiced far beyond the confines of the classroom. Still about the character of student life, these stories provide additional performance elements: costumes, stagings, and props. At Hamilton College, for instance, alumnus John Hudson Peck reminisced about the importance of math-oriented costumes for making his class *nulli secundus* (second to none). The ultimate student experience, for this lawyer-administrator, was not so much singing but rather, as Butlerian historians might anticipate, putting on a math show and impersonating young women.

Female impersonation has formed a core student experience for many young men, as educational historians and critics have noted. Recently, educational historians Margaret Nash, Danielle Mireles, and Amanda Scott-Williams have located Burials of Euclid in the history of "drag shows" on college campuses.[22] Earlier, archeologist Laurie Wilkie noted the ways female impersonation supported the late-century masculinities of California fraternities.[23] Critic Marjorie Garber more broadly argues that what she calls "transvestism" involved the conflation of anxieties about gender identity with anxieties about other identities—in Peck's case, the identity of "college student."[24] Informal performances of math communication similarly cemented masculine student identities through specific traditions of female impersonation.

When Peck addressed Hamilton College audiences in 1909, they likely expected him to tell them about the project of technical education, not about female impersonation. Though educated in liberal studies and experienced as a lawyer, he had been president of Rensselaer Polytechnic Institute from 1888 to 1901, and he had become a national advocate for engineering. He had organized technological exhibits for the Chicago World's Fair in 1893, he had represented educational interests at the New York State Constitutional Convention of 1894, and he had served as a trustee of the scientifically pathbreaking Troy Female Seminary.[25] Furthermore, because of his professional identity as a lawyer, Peck often spoke before groups of engineers about the connections between his educational background and their work. "It so happens that I am not an engineer, that is, I was not educated as an engineer," he explained to the Chicago meeting of the Western Society of Engineers in 1897.

> I am very fond of meeting with them. I have met a great many of them, and I always enjoy an assemblage of engineers. There is a different expression on that audience from any other that I ever face. There is more intensity, there is more directness, it is different, I cannot explain it exactly, I cannot describe it, but it is a different expression, and it is one that I am sure any man who is up-to-date, any man who loves his country, any man who loves his fellowmen, is delighted to see on any audience of American citizens.[26]

A technological enthusiast, Peck's words could not always keep up with his passions, and he developed a strong reputation as a public and political advocate for engineering, not for the women's clothes worn when burning and burying a calculus textbook.

Peck's college address began in the usual way. After all, forty-seven years of similar texts had established how Hamilton alumni were supposed to remember their student days. Peck started by noting the importance of having a living alumnus talk about his memories of student life a half century ago. No other histories, not even college histories, would capture his perspective, he explained. Though difficult and imperfect, his story would "try to waken echoes" of past youth, and he expected some of these "echoes," though familiar to him, to be "novel to newer lives" of younger alumni and current students. In short, Peck's speech tried to capture the character of his college class for the purpose of entertaining and educating the broader Hamilton community. The form of the speech remained, as it had been, a letter addressed to the Hamilton alumni as a whole. Throughout the nineteenth century, the "half-century annalist letter" marked a communal opportunity to celebrate the college, and it remains a distinctive Hamilton tradition to date. Just before Peck's, these "letters" had transformed from public texts published in national newspapers to distinguished speeches delivered before alumni audiences. Explicitly epideictic, they continued to praise the college for the purpose of fellow feeling and good publicity.[27]

Peck's first stories were not the "novel" ones that he promised; rather, they followed his initial visit to campus. He first came to the college with a brother and their father (Hamilton class of 1825). Both sons had attended a preparatory school, but growing up in Eastern New York State, they had no intention of going so far away for college. They thought it was just a "lark" to visit their father's alma mater with him, but it was much more. The beauty of the area, the friendliness of the people, and their father's unexpected jokes ("just not like father") all surprised the boys. When their father decided to call on an old friend, now a professor, they found the professor similarly good-humored and smiling. Brought to Professor Oren Root's house, the boys began to take the college entrance examination amid their elders' smiling faces. The Peck boys continued their visit by learning some of the history and lore of the college and surrounding area, including the reasons for the college museum and even some student pranks. The elders began to introduce them to other friendly young men who had decided to enroll there. When their father left for home, both boys stayed behind.[28] By remembering his first college visit, Peck followed a tradition of similar stories that established how college entrance depended on alumni connections, campus beauty, professorial good humor, and peer friendships. Though the personal details would differ from one person to another, his initial story was not unusual for the time, nor would it be very unusual today.

How did Peck start to talk about the importance of burning textbooks and impersonating young women? Like other Hamilton "annalists," he tried to relate not only the common stories but also the central character of the class of 1859, what he called "the class spirit." Despite their diversity, being anywhere between sixteen and twenty-seven years old at admission, the class came together through not only their required studies, Peck claimed, but also their shared social experiences. As freshmen, they treated sophomores with a mixture of respect and fun. Among themselves, they pooled their resources for food as well as sports and coursework. They were diligent but also healthy, combining intense work with intense exercise. By the end of their freshman year, Peck remembered, the class of 1859 had decided on a motto of *nulli secundus* (second to none), and they took the lead in the social life of the college: founding a college newspaper, joining fraternities, and establishing "smokers' and card clubs." Improving the college community "in sympathy and manliness," they also observed the usual rites of passage. They rang the bell to mark their advancement from "boyhood" to sophomores and therefore to college men. They did not lead the pack in their hazing of freshmen.[29] Instead, they most embraced their "sympathy and manliness," their sense of being nulli secundus, in burning and burying their calculus textbook.

And so the recognized advocate of engineering education, Peck, remembered how his class buried calculus and how they did it *best*. In a burst of eloquence that was somewhat unusual for him, he explained,

> At the close of our first term junior, we had a celebration and a jollification which were also far more. The mental athletics and polishing processes of pure mathematics for us were over. Old Calculus was 'off his job,' dead. Arrangements for his funeral had occupied a considerable portion of life among us. Obsequies began as examinations closed. A departure from College customs, the plan and its execution were immeasurably beyond the possibility of our origination in any

previous term. It was a frolic that showed transforming progress. The project had been carefully studied, and it was elaborate and dignified.[30]

Embracing the spirit of "nulli secundus," Peck explained, he and classmates constructed a large, complex celebration with many different components. A classmate, who Peck remembered as "Hall," wrote songs in Latin, English, and a combination of the two. Another classmate delivered a chapel eulogy that was "burlesque" but still respectful. A third, who was already training to be a minister, marched as the character Gen. E. Ration (generation), with others as Major Axis and Square Root. Another four dressed in women's white gowns, pretending to be the parodic Minnie Mum (minimum), Pollie Nomial (polynomial), Eva Lution (evolution), and Jennie Ration (generation). The underclassmen marched next to these young men in gowns "with a gallant devotion to it as if they had been in fact the young ladies they seemed to be."[31]

In fact, female impersonation was central to Peck's memories of the superiority of his class's Burial of Calculus. "Under starlight," he explained, "the illusion certainly was perfect and pretty, and the sobs quite moving. One small boy from the village did confide to his sister that a funny lady 'sword' at another funny lady who had stepped on her dress."[32] For Peck, such confusion established the amazing success of his class at achieving a certain measure of realism. After all, the "small boy" apparently thought he saw a "funny lady" acting oddly instead of a funny young man. After the class's procession, when the pallbearers Sir Face (surface), Sir Cumference (circumference), Sir Veying (surveying), Cy Cloid (cycloid), and Ab Scissa (abscissa) placed the coffin on the fire, the class gathered together to drink and continue sobbing and singing. Peck claimed that he was the one who donated the "urn" that the class used to gather the ashes of the textbook, now buried somewhere on the college grounds. He ended the story by saying, "The whole thing was a complete surprise to the College as well as a success and established the reputation of the class for initiative."[33] For Peck, the Burial of Calculus therefore represented the advancement, originality, cooperation, and fun of his class. Like the later academic rites of passage (public examinations, the inauguration of a new college president, prize exhibitions, and finally good-byes and commencement), the Burial of Calculus demonstrated the ways that the class came together as a social group and how they progressed (and processed) together. For Peck, it proved emblematic of the class spirit, that they were "nulli secundus," at least when it came to burning math books.

The math costumes, the prime figure of Peck's reminiscences, did have connections to the classroom and its lessons. As the earlier Burials of Euclid exposed geometry terminology, the new Burials of Calculus clarified the justifications for calculus and how it fit within particular visions for technical education. Like technical education as a whole, calculus rituals emphasized ideas of leadership, generalizing from the new careers of industrializing America. Their enactment emphasized the development of broad-based managerial skills—that is, the potential to convince groups of young men to follow orders and operate together. Seemingly borrowing the vision of gentlemanly leadership from Yale and elsewhere, technical education was nonetheless different. It grew out of the utopianism and republicanism of American technology, as historians John Kasson and Ruth Schwartz Cowan have recovered: the idea that industry could simultaneously cure the nation's ills and equalize social roles.[34] Yale's reports and rituals, with their class-based exclusivity and emphasis on traditional, classical learning, did not sit well with the developing worldview. Rather, at the time, new institutions took up the torch and flaming textbook.

At Hamilton, in short, the sense of superiority was ultimately entwined with the new calculus classes, which had been started by the kindly professor who appeared at the beginning of Peck's tale. In the eighteenth and nineteenth centuries, Hamilton had emerged as one of the three foremost institutions of men's higher education in New York State, along with Columbia and Union College. It owed its reputation to its age, atmosphere of religious tolerance, and close connections to older, "colonial" colleges: Harvard, Dartmouth, Princeton, and Yale. Though Yale's example had been central to its curriculum, Hamilton's professors also started to teach staples of technical education: agricultural chemistry, civil engineering, mineralogy, geology, and calculus. While Yale had the resources to establish an affiliated scientific school for such studies, Hamilton had to create a curriculum somewhere in between liberal studies and technical education.[35] Hamilton's professors argued that calculus was a technical subject that created leaders who could attain career success in both traditional and new jobs.

Professor Oren Root, Peck's father's friend, embodied a spirit of pedagogical compromise. Father to statesman Elihu Root and son of the previous Elihu Root, he was an 1833 graduate of Hamilton College, where he learned the liberal studies of Greek, Latin, and mathematics. After serving as a tutor and librarian, he taught for twelve years in New York academies, gradually building an interest in geology and a personal mineralogical collection. In 1849, he was asked to join the faculty in what he called "the Professorship of Mathematics &c." At Hamilton, he added a calculus requirement to the usual subjects of algebra, geometry, and trigonometry, and he also rotated through classes in astronomy, mineralogy, geology, Latin, history, and English. He relied on his collegiate training for his knowledge of ancient languages and basic mathematics, but the technical offerings emerged from his personal interests and collections. Root's lectures in conchology and mineralogy, for instance, combined his passions for the theoretical accounts of the history of the earth with practical applications to mining and mine administration.[36] As a Hamilton alumnus with an autodidact's technical education, Root introduced mathematics classes that similarly bridged historical and modern topics, as he argued calculus would.

Root's chosen textbook told students why his subject was important for their future lives. Elias Loomis's *Elements of Analytical Geometry and of the Differential and Integral Calculus* began with the statement that calculus was not just for industrial jobs but for "the mass of college students of average abilities" who each might follow the traditional paths and become a "physician," "divine," "jurist," or "statesman."[37] Loomis, like Root, had a complex educational background. A Yale graduate from the class of 1830, Loomis had served as a tutor under President Day before leaving to pursue advanced work in the sciences in Paris. On returning, he had taught for the Western Reserve College in Ohio before joining the mathematics faculty of the new University of the City of New York (now New York University).[38] Loomis's textbooks, like his academic appointments, remained embedded in technical education, though he also argued consistently that it did not upset the traditional liberal studies. His calculus textbook motivated problems in astronomy, physics, and engineering by focusing on the "functional" equations of conic sections. Still, he maintained that the primary purpose of calculus should not be its applications but its "discipline of the mental powers."[39] As such arguments led to the incorporation of calculus into liberal studies, they also mixed career expectations with broader educational values. In Root's calculus class, Peck's classmates learned these claims, and they continued to repeat them even in campus traditions.

The Burial of Calculus program (now saved in the Hamilton Library) confirms a connection between classroom practices and textbook burials. A small, locally printed, four-page pamphlet records the ways that Peck's class dressed as and then enacted calculus puns. As he imperfectly remembered, the celebration included the marshals "Gen. E. Rating" (generating), "Maj. Axis" (major axis), and "Esq. Root" (square root); the pallbearers "Sir Veying" (surveying), "Sir Cumference" (circumference), "Cy. Cloid" (cycloid), "Sir Face" (surface), "Hi. Perbola" (hyperbola), and "Ab. Scissa" (abscissa); and the "virgin" mourners "Poly Nomial" (polynomial), "Minnie Mum" (minimum), "Jennie Rashun" (generation), and "Eva Lotuion" (evolution). All these calculus puns relied on the central terminology from Loomis's textbook and Root's class. Also, as Peck related, the main activities involved a Latin song and mock eulogy in the chapel, a procession to the gravesite, and mathematical jokes at the gravesite. There, they took "a final (re)view of the deceased," in which they stomped on the text (saying they "went over" it), threw it above their heads (saying they "under-stood" it), and tore it to pieces (saying they "saw it through"). The "female" mourners burned the remaining papers on a fire, in accordance with the students' understanding of the ancient ritualistic role of "virgins." Meanwhile, the rest of the class sang in Latin-English doggerel. They gathered the ashes in an urn, and they buried it while singing a final song (a "dirge" in English).[40] The extensive funeral arrangements required extensive planning and coordination, and according to Peck, it made the class of 1859 feel like campus leaders.

The calculus terminology appeared in both students' costumes and also their songs. The Latin ode they sang in the chapel, for instance, contained the chorus "Functiones ejus nunc / Omnes sunt defuctae" (loosely translated as "His functions are *never* defunct"). In less obvious mode, some verses of the English dirge evoked Loomis's mathematics. The beginning of the song used the metaphor of smoke to discuss the confusion of the occasion, and the second version ended

> Aloft the smoke clouds roll,
> In spiral volumes roll,
> With radius vector infinite
> Outstretching from the pole.[41]

Such terminology harkened back to Loomis's emphasis on algebraic curves, and the end of the song followed calculus to the entrance to the land of the dead, where the paths of the students would never more intersect with that of calculus:

> In vain he'll flounder and roar,
> Whirled from the Stygian shore,
> For to his curves our course shall be
> Asymptote ever more.[42]

Ultimately, these songs, sung at the end of the class, served as a paradoxical mathematical review, reminding the juniors of their calculus knowledge while also indicating how such knowledge would be forevermore "asymptotic" to them. The characters and songs joined together in encouraging the students to embody their new knowledge even when they engaged in an elaborate ritual that, they sang, would distance them from it.

What were the students trying to say about their calculus requirements exactly? According to their costumes, their feelings remained intentionally ambiguous. After all, some students dressed in white and attempted to take on the appearance of women. As Nash, Mireles, and Scott-Williams have argued about campus drag generally, gender bending reinforced the complexity of the occasion.[43] The math characters already combined elements of liberal studies and technical education through modern names and classical inspiration. Furthermore, while these students distanced themselves from their character by insisting that they were neither mathematical nor female, they also wished to maintain the illusion. Within the smoke of their burning books, these students embodied intentionally ambiguous roles, as critic Marjorie Garber has argued about broader traditions of transvestism.[44] In Peck's case, the students mixed identities between male and female, between liberal and technical, and between loving and hating mathematics.

Overall, the Hamilton students were not second to none just because they practiced what they learned from their math classes. After all, Yale students had done that already; so had many students at many other colleges and schools throughout the country. Instead, their costumes' ambiguity (between earnest and mocking) meant they were the best. These men learned how to assert their school spirit by putting on a complicated mathematical show.

## Creating Mathematical Order at Bowdoin

In Burials of Euclid and Burials of Calculus, students practiced their math communication, ultimately putting on the characters and costumes inspired by classroom experiences. Elsewhere, their sense of learning math came through theatrical staging and blocking of similar math "funerals." Staging and blocking, in theater, are used almost synonymously to refer to the controlled movement of actors (and other performance elements) in order to create a scene. In Bowdoin's Burials of Mathematics, a sense of orderly movement came to the fore, showing the classroom connections between math and industrial management.

In the Bowdoin classroom of the time, Professor William Smyth instituted new calculus requirements, an attempt at mid-nineteenth-century curricular reform. Similar to Hamilton's, but more explicitly, Smyth's calculus built on a vision of America's scientific leadership, spelled out in his textbooks. Smyth based his algebra and trigonometry textbooks on the work of French authors Sylvestre François Lacroix, L. P. M. Bourdon, and Étienne Bézout.[45] His calculus also relied on French editions, and moreover, it contained numerous examples in the physical sciences from British sources: solving equations for motion in mechanics, calculating equilibriums and centers of gravity in hydrodynamics, and illustrating celestial motion in astronomy. He even motivated the whole study of calculus through a physics problem supposedly by Isaac Newton: "find the space through which a body, acted upon by gravity, will descend in a determinate time."[46] For Smyth, calculus provided not only the best route to the sciences but also the mathematical outgrowth of scientific investigation. The United States, he claimed, was particularly situated to take advantage of the best of calculus. A former British colony that borrowed from French classrooms, the United States already brought together the competing European traditions of calculus study, for Smyth. Combined with a scientific bent, Bowdoin's version of calculus could produce young men who could lead nationally and internationally in science and industry. Smyth emphasized that such classes would give his students industrial leadership worldwide. Teachers acknowledged the power of Smyth's arguments when they started to adopt his methods and textbooks throughout New England, from Bowdoin to Harvard and far beyond.

Staging emerged with utmost importance when Smyth's students began to burn and bury his books. In 1852 or early 1853, some Brunswick resident acquired a "Programme of Exercises" from Yale's November 1851 Burial of Euclid, and the students held a similar "Burial of Mathematics" on August 30, 1853. The celebration began, according to the pamphlet, at the mathematical recitation room at seven o'clock in the evening, and the group then processed to the chapel with the textbook "corpse." William Washburn acted as chief marshal, with the assistance of John A. Douglas, Charles F. Todd, Charles W. Smyth, and Andrew Ring Jr. The "Exercises at the Chapel" began with a voluntary performed by William M. Bartley and a prayer from the (mock) "chaplain," H. N. Merrill. Bartley then led the class in a song he composed, with words from J. L. Hatch. Singing "O Tyrones . . . vero! vero," they used its Latin words as a joking plea for truth to be visited on the underclassmen, the "Tyros." These themes appeared again in Wilson's eulogy and Hatch's elegy.[47] In including their names alongside their roles as "chief marshal" or "chaplain," Smyth's students implied some importance for their jobs at the mock funeral, for the ways they moved people: a parody of industrial leadership that was supposed to emerge from the classroom.

In laughing at authority figures, the Bowdoin students, unlike the Hamilton ones, also incorporated a set place for their professor and textbook author. In the "torch light procession" from the chapel to the grave, Professor Smyth was supposed to march in the second row, just behind the student marshals, and flanked by (mock) police.[48] He never did participate, but as the students jokingly remarked, they always found a volunteer to act his part. In the parade, the juniors performed a (dubious) desire that he would join them, but they controlled his stand-in's movements, just as they were supposed to control industrial settings. In the ceremonial display, they did not hide their purpose. A few rows behind the "professor," pallbearers held a coffin aloft, and their dirge imagined a future grave for Smyth's notable subject. Not just Smyth but also a personified calculus had a place of honor in the procession, and the juniors' specific spots (in front of everything else) made fun of curricular rationales. As they clearly learned, they were supposed to be leaders, choreographing others around them.

Bowdoin students mocked the supposedly managerial training by instilling a noticeable order into their proceedings. The procession, after all, included all figures in assigned places: chief marshal, assistants, professor, police, band, committee of arrangements, sexton, pallbearers, chaplain, eulogist, elegist, college classes, and even "Citizens Generally." Moreover, in their pamphlet, they marked their route clearly: "At 8 1–2 o'clock the Procession will move down Park Row to Pleasant Street, through Pleasant to Union, down Union to Mill, through Mill to Main, up Main to School, through School to Federal, down Federal to Mason, through Mason to Main, up Main to School, through School to Federal, up Federal to Bath street, through Bath, Main street and Professors' Row to the burial ground in the rear of the Colleges, where it will form an Ellipse round the grave."[49] These detailed directions were also not the only ones printed. After the short instance of prayer and song at the "gravesite," students and all were supposed to follow an abbreviated path back to the site of the mathematics recitation room, where the crowd would disband. Bowdoin's Burial of Mathematics therefore managed to create a certain routine out of the unruly practice of burning books. Bowdoin students mocked not only the interconnection of leadership qualities with their calculus classes but also the ways that these related to, in their words, the bold, capitalized "**ORDER.**"[50] It was all a matter of strict staging.

The conflation of calculus and leadership, in fact, encouraged the inclusion of Bowdoin student names, which resulted in the whole printed document appearing significantly different from earlier ones. Their inspiration, Yale's Programme of Exercises, had served as a record of the changed lyrics to common songs. Every page was filled with what Root and Lombard called "college poetry": a Latin version of the Scottish "Bruce's Address" on page 1, a version of the nineteenth-century minstrel song "Susanna" on pages 2 and 3, and a rewritten "Auld Lang Syne" on page 4. The format identified the source for each song without any indication of the lyrics' authors. The only names that appeared were clearly parodic: "Ghorr Rheigh" ("gory") gave the initial oration, and "Strong-in-the-Faith Redfield" (a character they constructed) delivered the funeral sermon. Yale's name itself appeared nowhere (see figure 1). In fact, Bowdoin's archives mistakenly labeled Yale's program as one of their own, not noticing the presence of the same songs in *Songs of Yale*.[51]

Bowdoin's Burial of Mathematics, by contrast, read as a litany of student names, and it continued to conflate students' management of the procession with leadership in the college. Eight distinct names appeared on the first page, identifying them as marshals, an organist, a chaplain, and a lyricist. Between songs and prayers, the second contained the names of the eulogist, the elegist, and another lyricist. The order of the procession identified nine more, and a final lyricist appeared on the fourth page. In sum, the program included the names of nineteen separate students, more than half of the total class of thirty-seven. Their names joined the announcement of collegiate pride that took up nearly half of the front page. There, a modern type proclaimed the "BURIAL OF MATHEMATICS." Below, decorative type most often found on contemporary posters, displays, and advertisements identified the participants

# BURIAL OF EUCLID.

## November, 1851.

## PROGRAMME OF EXERCISES.

1. OVERTURE—*From " Somnambula."*
2. CARMEN INTRODUCENS *CANTUS—Bruce's Address.*

### I.

Oi ! Oh ! deplorantes nos
*Bacchanalem* comitem,
Atra sors persequitur,
   Lacrymæ fluunt ;
Salsas lacrymas damus
In recentem tumulum
Euclid enim carus est,
   Recordari nunc.

### II.

Oppidani, ha ! ha ! ha !
Tutoresque, O ! O ! O !
Verbis tam innoxiis
   Quantum terrent nos !
*College leges* expellent
Omnes funerarios ;
*College leges*, ha ! ha ! ha!
   Quantum terrent nos !

### III.

Eia ! age ! juvenes !
Sophomores ! surgite !
Vos præbete strenuos
   Vaste ruite !
Oppidanos fortiter
Fustibus abigite ;
Facultatem clara vox
   Coget fugere.

### IV.

Seniores, salvete !
Veteres in scelere,
Nobis vester adventus
   Est gratissimus.

Juniores, salvete !
Dignitatis viridis,
Gaudemus conspicere
   Hic præsentes vos.

### V.

Vos tirones, salvete !
Vos, qui matribus caris
Estis audientes non
   Ah ! quam scelesti !
Sed egressi foribus,
Illis non sinentibus,
Elephantem visere
   Huc venistis nunc.

### VI.

Pater Euclid cum abit
Dixit, nobis flentibus,
" Ne me lamentamini,
   Hilarescite.
Biennale referet
Umbram cari Euclidi."
Luctum ergo tollite ·
   Luctum tollite.

### VII.

Euclid nunc humabimus
Splendide ac fortiter ;
Præses dicat quidlibet
   Thacher Kinneque
James, qui magnum nasum fert,
James, qui stans collare fert,
Omnes tales ; pulchri sunt,
   Verum possunt non.

FIG. 1  Front page of "Burial of Euclid, November 1851, Programme of Exercises" (Yale). Courtesy of the George J. Mitchell Department of Special Collections and Archives, Bowdoin College Library, Brunswick, Maine.

and institution: "the JUNIOR CLASS of BOWDOIN COLLEGE" (see figure 2).[52] In loudly declaring their institutional affiliation, the Bowdoin students claimed that their roles in the student production approached the managerial task of running their college. Though literary societies held a similar place in allowing students to take charge of their curriculum and social rules, as other historians have shown, the textbook burials did not have the official endorsement of Bowdoin's societies.[53] Rather, as the program's boldface "JUNIOR CLASS" made clear, these events ostensibly represented the whole year. It allowed them to assert that their collective body, not calculus or Smyth, really ran the show.

The student songs similarly commented on the power of knowing calculus. In August 1854, the juniors prefaced their "torch light procession" with a version of "Auld Lang Syne" modified for the occasion. Later, they sang a version of the hymn "China," also known as "Jesus Loves Me, This I Know," and their version encouraged a contest between their love of Jesus and hatred of calculus.[54] They made the difficulties of the class clear through their irreverent second and third verses, rife with college slang with possible sexual connotation:

Old Calculus has screwed us hard,
Has screwed us hard and sore,
I would he had a worthy bard,
To sing his praises more.

He took the strongest of the class,
And brought them to their knees,
And then we found too late, alas!
Prof. Smyth was hard to please.[55]

Singing their displeasure before a town-gown audience, they noted how the (exaggerated) calculus requirements served as another rite of passage for them. By expressing their irreverence toward their college, the juniors proclaimed their assumed superiority.

Overall, the emphasis on the work of staging a student event communicated a certain vision of the career preparation of calculus. Bowdoin's students, like their counterparts at Hamilton, learned about the interconnections between calculus requirements, math applications, and industrial leadership. A liberal education (as we saw in chapter 1) was fundamentally a matter of engaged citizenship, producing educated young men who could return to their communities and advise on a huge range of issues. Along with Smyth's arguments for new calculus requirements, Bowdoin's notion of civic responsibility had to transform. As the student reactions demonstrated, the emerging pedagogical vision became a matter of seeing opportunities to cooperate in the production

# BURIAL OF MATHEMATICS

## ~ BY THE ~

## JUNIOR CLASS

### ~ OF ~

## BOWDOIN COLLEGE,

### AUGUST 30, 1853.

The JUNIOR CLASS will meet at the Math. Rec. Room at 7 o'clock P. M., and will thence proceed with the Corpse to the Chapel.

CHIEF MARSHAL.

### WM. D. WASHBURN.

ASSISTANT MARSHALS.

| JOHN A. DOUGLASS, | CHARLES F. TODD, |
| CHARLES W. SMYTH, | ANDREW RING, JR. |

## EXERCISES AT THE CHAPEL.

| Voluntary on the Organ, | WM. M. BARTLEY. |
| Prayer by the Chaplain, | H. N. MERRILL. |

### CANTUS.

Numeri a BARTLEY.  Verba a J. L. HATCH.

Ut conticeatis omnes obsecratum est—
Voculas—O Tyrones—verecundi comprimete!
Fiat altum silentium!
Extenditur enim in conspectu mortuus—
   Vero—vero!
In conspectu jacet informe cadaver!

FIG. 2 Front page of "Burial of Mathematics by the Junior Class of Bowdoin College, August 30, 1853." Courtesy of the George J. Mitchell Department of Special Collections and Archives, Bowdoin College Library, Brunswick, Maine.

of a vast, interrelated machine. The funeral program, after all, contained student names alongside administrative, managerial roles. The students, in performing these jobs, showed that they had the skills necessary for controlling a complex operation. They aspired to industrial leadership not through performance per se but through role-playing management and administration.

According to records of these students' later employment, participating in the Burial of Mathematics actually reinforced classroom lessons about math's role in career preparation. In other words, the students did not simply hate calculus, even when they said so. While some of the nineteen listed became lawyers, members of the clergy, or in one case, a music dealer, nearly a third pursued careers in science and industry. The chief marshal, an assistant marshal, and a member of the arrangements committee all fulfilled the classroom visions and became managers of factories in New England and the Midwest. Another assistant marshal and a pallbearer became physicians, settling in Massachusetts and California before joining the war effort in the 1860s. While the third assistant marshal died before graduating, the fourth became a mathematics professor at Catawba College in North Carolina.[56] The future professions of all the students (especially of the professor-to-be) demonstrated how participation in the burial ceremony allowed an opportunity to celebrate calculus and the possibilities that it introduced.

What exactly about math communication was practiced in Bowdoin's Burials of Mathematics? As Smyth said in the classroom, calculus provided a way for Americans to engage the sciences and compete internationally. His students built on these claims through their staging of a public performance, gracing the streets of Brunswick in the middle of the night and probably annoying many, many people. Viewed generally, students' mocking arguments repeated their professor's earnest claims: calculus made them industrial leaders by giving them managerial skills, especially ordering people's movements. As graduates, they could (and did) take a wide variety of positions in America's emerging technological society. The blocking/staging of their math "funerals" showed how much of the message they understood and ultimately practiced.

## Competing Mathematically at Bates

Increasingly, practicing math communication became a sign of college identity. When professors, presidents, or other college officials celebrated math "funerals," it was only through reminiscences of their own student days. In fact, these events were increasingly prohibited well into the twentieth century. Still, from actual students' perspectives, these events mattered for constructing their identity as math students—perhaps a mathematical version of Horowitz's college men[57]—which occurred especially in response to stories of similar traditions elsewhere. In the case of Bates's Burials of Anna Lytics, these rehearsals for math communication proclaimed college status particularly in dialogue with stories of other students at other colleges.

Bates's programs, in particular, clearly reacted to the ones at Bowdoin. Bates's origins as Freewill Baptist had roots in the beliefs of its first president, Oren Burbank Cheney, who argued before the Maine state legislature that Bowdoin and Colby could not satisfy all the higher education needs of the state. With a small faculty in the classics and moral philosophy, Cheney's school received a charter in the mid-1850s, and it experienced expanded offerings, expanded buildings, and expanded purposes during and after the Civil War. Unlike other colleges in the state, Bates quickly offered higher education to African Americans and to women, reflecting Cheney's beliefs.[58] The Burials of Anna Lytics, as practiced at Bates, emphasized societal divisions instead, providing a striking counternarrative to Cheney's public proclamations. For the select Bates students who participated, these math funerals demonstrated an attempt to assert Bates as a college, with cultural presumptions of white, male education.

In printing math programs, Bates students participated in what turned out to be a national phenomenon. Starting at Yale, the tradition of "burying Euclid" had spread like wildfire to at least fifteen American colleges by the 1890s.[59] In these rituals, students mocked and ultimately practiced the lessons learned in the classroom. Through allusions to common terminology, participants indirectly embodied math knowledge. They also more literally embodied the knowledge through what became the festival's usual characters: Euclid, Geo. Metry, and especially Anna Lytics. By burying their math books, especially creating funeral programs for the character of Anna Lytics, Bates students therefore symbolically joined the ranks of college students. The act of construction and destruction did not just matter for math, locally and nationally. Beyond Bates, these student traditions always had included allusions to other college subjects: mainly Latin and Greek, with occasional references to natural history, English literature, logic, economics, geography, and even rhetoric and composition. Taken in full, Bates's Burials of Anna Lytics implicitly and subtly argued that Bates was a college, particularly one on par with the older and more established Bowdoin College. Through attention to design, song, and intertextuality, Bates students used math programs to assert they were college students too.

Bates students borrowed many design features from the earlier Bowdoin programs. Bowdoin's programs had already gone through a change. As we saw before, Bowdoin's Burials of Mathematics had read as a litany of student names, conflating students' management of the proceedings with mock leadership roles. Though songs appeared in ancient languages, the text generally communicated the roles and activities in English.[60] Later in the century, after the Civil War, Bowdoin's Burials of Anna Lytics emphasized what an alumnus remembered as the "formal" qualities of the affair. For him, the printed program was a way to excuse a practice designed to show how "some students . . . did not enjoy that particular branch" of mathematics. It lent a serious quality to the proceedings through "a regular programme printed in black type and with a heavy black border," including the participants' names "given in a peculiar classic Latin," as well as "literary exercises" of processions, speeches, and songs.[61] In other words, the postbellum program was "regular" and "literary," turning a mock funeral into something more like an exercise in Latin composition.

Bates's Burials of Anna Lytics included the same design features in the same order, demonstrating their collegiate identity through implicit comparison with Bowdoin's. Their programs similarly consisted of lists of similar roles in "peculiar . . . Latin": *Laudatio*, *Elegia*, *Cantus a Cantoribus*, all the way to *Ceteri Ploratores*. Between songs and roles, an entire page indicated the correct processional order, what students at Bowdoin and Bates both called the "ordo." Bates students, following Bowdoin, favored a morbid header of a simple skull-and-crossbones design, and they borrowed the generic black border too (see figure 3).[62] Implicitly, each design feature linked the Bates tradition with an older, established ritual associated with college life. More explicitly, when the Bates students listed their identity in large type on the first page, they borrowed the Bowdoin title. Instead of "Humatio Annae Lyticae in Collegio Bowdoinensi," theirs was "Humatio Annae Lyticae in Collegio Batesini." The title conflated their Burials of Anna Lytics with their status as students of Bates, clearly represented as a college here.

These arguments mattered because Bates's college status was not assured at the time; even the name "Bates College" was new. The founder of the college, Cheney, had attended a seminary named Parsonfield, and after he pursued a course of studies at the more liberal Dartmouth College, Cheney returned to Parsonfield to create an atmosphere friendly to the abolitionist cause. He succeeded, and by the early 1850s, Parsonfield had begun to house not only pupils but also fugitive slaves. Unfortunately, the situation did not last long. In 1853, the school burned down, perhaps because a local wanted to murder the residents for their political leanings. After a period of mourning for the five victims, Cheney campaigned for a replacement institution, though one focusing on more advanced studies, and by 1855, he convinced the Maine state legislature to allow a charter for the Maine State Seminary. A period of intense fundraising commenced, and after a remarkably large donation from Boston businessman Benjamin Bates, the seminary became Bates College in 1863–1864. The name change did not assure its academic reputation, however. Cheney had argued that the institution should be open to women and African Americans, and when Mary Mitchell and Henry Wilkin Chandler graduated in the 1860s and 1870s, some students at Bowdoin publicized their derogatory

# HUMATIO ANNÆ LYTICÆ

— IN —

## COLLEGIO BATESINI,

IN QUARTO DECIMO DIE ANTE CALENDAS QUINTILLES,

### MDCCCLXXIX.,

#### CELEBRABITUR.

## A N N A.

## '81.

Reliquiæ defunctæ videri ab amicis et multitudine in loco
Mathematico prima vigilia noctis poterunt; unde
ad LIMINEM TEMPLI ferebuntur.

FIG. 3 Front page of "Humatio Annae Lyticae in Collegio Batesini" (1879). Courtesy of the Edmund S. Muskie Archives and Special Collections Library, Bates College, Lewiston, Maine.

views, saying women and African Americans could not really attend college; they could only go to "school" at best.[63] Though such changes in the student body were common in postbellum America, as educational historian Michael David Cohen has shown, so was official and unofficial discrimination from white, male students and faculty.[64] By the 1880s, Bates's academic reputation had become tied up in such national and local debates, and as in many other institutions, Bates's students repeatedly asserted that they did not attend "school"; they attended "college."

The funeral programs provided a particularly striking opportunity for Bates students to argue for their college identity through song. Singing itself was becoming a recognizably collegiate activity, as historian James Lloyd Winstead has argued, and songs had become a primary component of textbook funerals as early as 1850s Yale.[65] Still, the mock funeral songs had served different purposes at different times and places. Students wrote new lyrics to common tunes, and their compositions emphasized specific information about their particular textbook, their campus's geography, and even their college's mission. Even the chosen tunes reflected different preferences. Where students at antebellum Bowdoin modified Protestant hymn tunes, such as "Jesus Loves Me, This I Know," Bowdoin students after the war favored the newer battlefield tunes, such as the "Battle Hymn of the Republic."[66] The songs used in the Bates programs ranged from folk tunes ("Auld Lang Syne" and "My Last Cigar") to ones we might think of as children's songs ("My Grandfather's Clock," "My Darling Clementine," and "My Bonnie Lies over the Ocean").[67] Still, they nearly all shared a common source: books of college songs. Following the popularity of *Songs of Yale*, collections of intercollegiate songs started to appear increasingly by the 1880s. The songs used in Bates's mock funerals, for instance, came from two such volumes: *The American College Song Book* and *College Songs*.[68] These volumes gave consumers the opportunity to sing like the students at Harvard, Yale, and Amherst, and they (in part) made possible the popularization of glee clubs and college choirs well into the twentieth century. The students at Bates used recognizably collegiate songs, inserted into a recognizably collegiate tradition, to make the argument that they too were college students.

Their argument did not last long. President Cheney did recognize the sexism and male privilege that the funeral programs implied. By 1890, he and the faculty established what the students nicknamed the "Anna law": it was no longer permitted to participate in the tradition of burying Anna Lytics.[69] As in the case of Bates's new fraternities, modeled on Bowdoin's, African American students had been practically excluded from the Burials of Anna Lytics, as had women. The role of women, in particular, was particularly worrying in Bates's burials. White, male students at Bowdoin could be viewed as participating in the national anxieties about the feminization of college education, especially of math classrooms. At Bates, as at Bowdoin, the choice of the name Anna seemed to warn educated women that their mathematical knowledge could lead to their burning, at least in a metaphorical sense. But it was one thing to joke about a woman's death at Bowdoin College, which remained all-male. At Bates, which already had a reputation for coeducation, the jokes seemed to express sexist anger or resentment, accompanied by violent threats. When the faculty began to enforce the "Anna law," suspending all participants throughout the early 1890s, the Bates student newspaper protested that the jokes should not be taken too seriously. Yet even these editorials replicated the conflation of white, male privilege and college education. It was a matter of "proud fathers" who "will not send their sons to a college from which they are sure to be suspended."[70] Though the Burials of Anna Lytics slowly died out in the last years of Cheney's presidency, such connotations continued to follow the attempts at student humor well into the twentieth century. The Bates faculty seemed to understand the implications of asserting too forcefully that Bates was a college, especially one modeled on older, white, male institutions.

In these programs, as in the broader events, what did the students think they were doing? Because these programs were designed to be ephemeral, there are not many sources about their construction. Though there are scattered alumni recollections from Bowdoin, as well as many other colleges, I have not found similar records at Bates. The closest indication is the copy of an Anna speech preserved in a student's file of college writing.[71] Between a graduation speech and essays about European history and politics, Charles Nichols saved the odd document. Ending "let us differentiate the infinitesimal elements of her inexhaustible, imperishable, incorruptible and unmeasurable [sic] ostentatious conglomerations," it seems an exercise in verbose language, exaggerating tropes of public speaking for comic effect.[72] Its presence in his file seems to communicate an understanding of the work that comes with creating an English composition, especially such a humorously overwrought one. Perhaps he wished the whole file to be a joke of the archives, as it begins not with his essays but with those of a female classmate, Margaret Knowles. Nichols's portfolio raises more questions than it answers.

Articles in the student newspaper served to support the view that the Bates students understood the ways their math funerals argued for their collegiate status. Founded in the early 1870s, the Bates Student included notices about local events, along with poems, stories, and news from "Other Colleges." The last section expanded from short descriptions of sports matches between Bates, Bowdoin, and Colby (in 1879) to longer features about life at Harvard, Yale, Brown, the University of Wisconsin, Nebraska, Pennsylvania, Chicago, Stanford, Cornell, and about a dozen others (in 1893). Reports about the Burials of Anna Lytics regularly appeared as long as the tradition existed, and their placement seemed to communicate the implicit expectation that their news could be copied into other student papers. In printing articles about their printed programs, the *Bates Student* served to locate their activities in what the student editors called the "College World."[73]

As time went on and these informal performances gained more components, math funerals around the country became not just about repeating what students learned in the classroom. The characters, costumes, staging, and programs pointed back to the math classroom, as always, but they also pointed outward, to other colleges where similar events could be found. As we will see in chapter 4, math plays—actual attempts at math theater—were similarly, self-consciously about college life. As explored here, math communication was practiced through mock "funerals," through occasions for burning/burying books. Such events were for a time synonymous with student life: times when actual students generalized about the whole of college identity from the vantage of the math classroom.

What was being rehearsed when actual students practiced math communication? In these historical cases, college students repeated lessons learned in their math classrooms, including particularities of terminology. Moreover, students used these performances to show off, compete with others, control the movement of people, and assert their institution as a college before local and national communities. These seemingly unrelated components were part of their math classes too. Display and competition entered the classroom through math drills and ideals of mental discipline, as in chapter 1. Controlling people and asserting (college) status came from math's curricular reforms of the day: in part through the importance of math for industrial management. Taken together, these historical cases should encourage us to rethink actual students' reactions to classes, especially when they seem destructive. There may be more of the classroom in today's pranks too.

These cases begin to tell us about the usual set pieces of math communication as well. It is not just about characters, costumes, staging, and programs of informal performances: it is also about the usual features of the math classroom—the ones often called "technical" themselves. The blackboard has come to the fore (literally) as the instructional technology of math education and technical education broadly. Math communication (and broadly technical communication) has depended on the presence of such technologies and developed around them, circling those technologies just as students circled teachers and textbooks' mock "graves."

# 3

# How Math Anxiety
# Has Developed from
# Classroom Tech

• • • • • • • • • • • • • • • • • • • • • • •

The blackboard has appeared already in this book, though it might not have seemed noteworthy. It is featured in relation to the Conic Sections Rebellion in chapter 1, and in chapter 2, Yale students mentioned it in one instance of the Burial of Euclid. Cobbled together with the introduction's brief mentions, the blackboard seems a material artifact, an instrument of pedagogical display, a site of tacit skill, and a symbol of a lone genius. Epitomizing classroom technology, the blackboard has represented the performance of teaching and learning. This chapter draws attention to the blackboard, analyzing some of the unspoken rules of oral communication associated with learning math at the board. It does so principally through looking at the performance of math communication at a time in the history of classroom technology when the blackboard was new.

The blackboard has received considerable attention in recent years as part of the larger project of exposing the material tools of science education and scientific investigation. Sociologist Donald MacKenzie and historian Michael Barany have investigated the ways that blackboards are constitutive of mathematical communities and mathematical research, even in the twenty-first century, when many classrooms are replacing them with newer pedagogical technologies.[1] As part of the broader project of the Sociology of Scientific Knowledge (as mentioned in the introduction), MacKenzie and Barany have sought to analyze the production of new mathematical knowledge not as the provenance of individual inspiration and ability but as the material product of human groups. In doing so, they have focused on the performances of human actors in research seminars, including principally their talk at the board and with the board. My analyses supplement theirs in showing how certain conventions have developed for boardwork in classroom contexts.

Connections to performance similarly have punctuated the blackboard's historiography. Math historians Peggy Kidwell, Amy Ackerberg-Hastings, and David Roberts have compiled a definitive overview of the blackboard's history with careful attention to its periodization. After noting that the blackboard's first uses in the United States remain unknown, they have proposed two "broad periods": before 1850, when certain mathematics teachers in the Northeast and Central Atlantic states used it as a supplement to students' slates, as an alternative to textbooks, and/or as a pedagogical device teachers could create themselves (using specific paints, etc.), and after 1850, when trends in building, transportation, education, and economics created national and regional markets for school supplies.[2] Using objects from the Smithsonian collections as representative, Kidwell, Ackerberg-Hastings, and Roberts have illustrated the first period with a photograph of a blackboard from the Glebe Schoolhouse in New Hampshire: pieces of wood painted black and with chalked exercises from the instructional manual *The Black Board in the Primary School* (1841). For the second period, they have drawn attention to a post-1857 globe covered in slating (for spherical geometry and/or geography) and a 1920s advertisement for a commercial blackboard available in green or black hypoplate.[3] Throughout, they have noted the changing expectations for student interactions with these material objects, whether performing mathematics at the front of a classroom, at a teacher's desk, or along the sides of the room. Their work now allows for greater attention to the ramifications of blackboard technologies for student communication and also student anxiety.

Math anxiety per se has a very short history. Often considered a version of performance anxiety and/or test anxiety, math anxiety depends on the framework of psychological anxiety, which emerged in the United States in the 1880s–1890s and was popularized through translations of Sigmund Freud in the 1920–1930s. Though based on an earlier medieval concept that was equivalent to worry (specifically, worry over an uncertain future), modern anxiety depends on a medical understanding of certain physical signs—rapid heartbeat, tense muscles, and/or shortness of breath—associated with what the *Oxford English Dictionary* calls "inappropriate or excessive apprehension or fear."[4] In other words, anxiety is not just about worry over particular situations; there must be uncomfortable bodily effects, something that could be deemed "inappropriate or excessive" from a medical standpoint. Experiences have been labeled "math anxiety" when they have become a problem, beyond the norm. Still, incidents of math anxiety have seemed to increase in recent decades, and books about math communication have been calling for techniques to "conquer" and "overcome" it since *math anxiety* was popularized as an alternative to *math avoidance* in the late 1970s and 1980s.[5]

Emphasizing math communication's connections to performance, I investigate the dimensions of stage fright in math anxiety. *Stage fright*, though similarly medicalized today, is much older and more common in Anglo-American contexts. *Fright* itself has an Old English analog, *fryhto*, which means fear and can be found in manuscripts from the 800s CE.[6] In the thousand years from *fryhto* to medicalized anxiety, fright has shifted meaning, referring to particular instances of fear: ones that are sudden or especially strong. Stage fright today can be associated with psychological anxiety disorders and is often described as a medical event. Characterized by worry, stage fright involves certain physical signs before or during a performance: rapid heartbeat, tense muscles, and shortness of breath, along with possible stuttering, tremors, tics, dryness, sweating, dizziness, nausea, and other signs of discomfort. Stage fright has had a resurgence in recent years, particularly as professional performers admit their struggles with the disorder.[7] Nearly synonymous with *performance anxiety*, stage fright has come into the classroom too. Students can experience stage fright when giving speeches, when presenting projects, or whenever they feel performance is required.[8] In math classes, stage fright is—and has been—usually about boardwork.

Looking at the connection between stage fright and learning math makes explicit certain rules for math communication. Though the classroom version of stage fright has had to do with certain defined tasks of oral communication—speaking about prescribed topics (in math) around and with particular technologies (the blackboard)—the rules might not have been so defined. Researchers in technical communication have noticed the ways that technologies mediate the experiences of public speaking, particularly paying attention to the understood (though often tacit) rules for PowerPoint.[9] Similarly, there have been certain social rules (that have developed over time) for how to interact with a blackboard, though they are often not stated explicitly. Looking back to moments of obedience and disobedience from the blackboard's first period of use can serve to recover the rules for speaking mathematically.

Certain historical cases emphasize the pedagogical technology involved in learning mathematics. Beginning with a short history of the blackboard, focusing on its early uses at the U.S. Military Academy at West Point, I will build an argument for how the military context has shaped the communicative rules surrounding boardwork.[10] Still, in line with the complexities of technology transfer, the blackboard has needed to have different rules when adopted beyond the military. A combination of mental discipline and blackboard policies, I argue, prompted the conic sections rebellions at Yale, the most famous example of "inappropriate or excessive" stage fright involving the blackboard's use. Pointing us toward chapter 4 and looking back to chapter 2, this chapter ends with a mid-nineteenth-century Yale lithograph that represents math on a proscenium stage a generation after the Conic Sections Rebellion and just into the second period of American blackboard use. Emphasizing the performative dimensions of math communication, stage fright acts as a historical analog of math anxiety, supported and surrounded by certain classroom technologies.

## The Blackboard Rules: The Military Academy

The U.S. Military Academy at West Point fundamentally shaped American math education, in large part because of the blackboard. Blackboards had existed in certain forms before the military academy's founding, particularly in other countries. Even in the United States, handheld slates and chalk were not uncommon before the Revolutionary War. However, large writing surfaces, so large as to fit on the side of a room, were extremely rare. Through a partnership with the French schools for engineering and military training, the military academy began importing these classroom technologies in its early years. By the 1820s, its professors were writing textbooks that strongly encouraged the adoption of blackboards, and these books proved widely popular and persuasive. When the military academy's graduates went on to become teachers at myriad schools, academies, and colleges throughout the country, they too requested blackboards. Educational journals, magazines, and societies advertised their benefits until they became ubiquitous.[11] Throughout, combining ideals of military life, engineering, and French education, the blackboard developed certain conventions in the United States around distinctively student-centered performance.

In its early American uses, the blackboard combined scientific education with explicit leadership training for the army and navy. As historian Alex Roland has noted, the U.S. Military Academy arose from bipartisan plans for a national university. Federalists, inspired by the Roman god of war, proposed a "University of Mars" with constant applications to the betterment of military training, and Thomas Jefferson and his supporters emphasized the scientific advancements that could come from such an institution.[12] Though Federalists and Jeffersonians disagreed about many things, they did agree about the ultimate purpose of the academy: to train military officers to lead and serve their country. Still, the best way to provide such an education remained under discussion for years, even during the hardships of the War of 1812, when the professors declared many of the cadets "graduated" because they were needed in military service. The curriculum did begin to solidify in the years after that war, and Superintendent Sylvanus Thayer encouraged a compromise between the original vision of the Federalists and the Jeffersonians. The U.S. Military

Academy, Thayer argued, should combine the French schools for engineering and for officer training (the École Polytechnique and the École Spéciale Militaire), teaching military leadership through the sciences.[13] Once adopted, the blackboard epitomized martial connections. As Christopher Phillips has argued, it expanded cadets' rules of behavior to include rules about how to draw, how to act, and how to speak.[14] The steady hand and firm gesture would communicate the cool mind of the officer-to-be. In other words, mediated math communication—following the rules for speaking in front of a board—became a sign of leadership ability, particularly in the hybrid military-scientific curriculum of West Point.

Broadly, West Point's leaders emphasized how French exemplars made cadets better officer-engineers. During the War of 1812, Thayer had visited France as a diplomat and spy, and he brought back a deep affinity for the French language and direct experience of the Écoles. He also brought back French professors, particularly in engineering. The most entertaining and inscrutable proved to be Claude Crozet, a young graduate of the École Polytechnique who also served as an officer under Napoleon Bonaparte. When decorated graduates, such as John H. B. Latrobe, wrote "reminiscences" of their cadet years, Crozet became a leading character. "There are persons whose appearance is never effaced from the memory," Latrobe began. "Of this class was the Professor of the Art of Engineering, Colonel Claude Crozet, tall, somewhat heavily-built man, not as straight, perhaps, as a cadet drillmaster would have made him, of dark complexion, black hair and eyebrows, deep-set eyes, remarkable for their keen and bright expression, a firm mouth and square chin, a rapid speech and strong French accent." After establishing Crozet's distinctly "French" appearance, Latrobe explained his teaching style: "He had been an *engineer officer* under Napoleon at the battle of Wagram and elsewhere, and the anecdotes with which he illustrated his teaching were far more interesting than the 'Science of War and Fortification,' which was the name of our text book at the time."[15] Embodying the cultural power of French styles for Americans, Crozet connected the institution to French engineering, scholarship, and military prowess. Additionally, through moving "far" beyond the textbook, he encouraged each cadet to imagine how he too could be an "engineer officer" of talent and renown.

The blackboard was similarly a recognizably French import. Earlier, as Kidwell and colleagues have noted, the word "blackboard" just meant a "wall painted black":[16] for instance, a tutor at Queens College (now Rutgers University) suggested a "blackboard" of this kind in 1779.[17] Similarly, the 1809 textbook *Arithmetic Made Easy for Children* recommended a "Black Board," by which it meant a board "painted or stained with ink" and mounted to the wall. Advocates of primary schooling in England had recommended that a (painted) blackboard be used to display white letters (i.e., the alphabet), but

*Arithmetic Made Easy* suggested chalk so that math problems could be worked out and then erased.[18] But French immigrants brought expectations for the blackboard with them. In 1814, a Catholic priest surprised his students when he used a blackboard at a Boston math school. According to one pupil (who would later become the area's biggest blackboard advocate), the priest's blackboard was not just an ample writing surface; it had "lumps of chalk on a ledge below, and cloths hanging at either side."[19] In other words, beyond a black wall, the device had extra components that made for easier writing and erasing. Under Crozet, West Point's uses of the blackboard became associated not only with French but also with math reforms.

According to cadet memoirs, Crozet introduced the blackboard in order to improve the military academy's math classes. "The surprise of the French engineer, instructed in the Polytechnique," remembered an anonymous cadet, "may well be imagined when he commenced giving his class certain problems and instructions which not one of them could comprehend and perform."[20] Though Crozet's main appointment remained in the "Art of Engineering," he began to take over more of the mathematical preparations too. Disgusted with the basic classes from surveyor Andrew Ellicott, mockingly called Mr. "Infinite Series," Crozet started to include introductions to geometry at the beginning of his engineering classes.[21] In order to aid these lessons, Crozet ordered a local carpenter to mount a large board on his classroom wall and paint it black: his blackboard.

Through his geometry for engineers, Crozet introduced not only a blackboard but also French descriptive geometry, including Pascal's theorem. An abstract statement having to do with conic sections (any shape produced when a plane intersects a cone), Pascal's theorem states, "If six arbitrary points are chosen on a conic (which may be an ellipse, parabola or hyperbola in an appropriate affine plane) and joined by line segments in any order to form a hexagon, then the three pairs of opposite sides of the hexagon (extended if necessary) meet in three points that lie on a straight line."[22] Given its formulation, what was the importance of Pascal's theorem for engineering? Conic sections, in general, helped engineers build tunnels, suspension bridges, arches, and certain retaining walls, since all relied on curved lines and surfaces. A specific rule like Pascal's theorem could provide a way to check a design or, under certain circumstances, calculate the stresses on a proposed or existing structure.[23] Pascal's theorem, French geometry, and engineering anecdotes quickly became staples of a distinctly West Point math, popularized through the blackboard.

Crozet, in fact, did not have an interest in popularizing these methods, but Ellicott's successor did. Charles Davies, according to Latrobe's reminiscences, "was a remarkable man."[24] Raised in a rural Connecticut county, he unexpectedly helped the army during an early battle of the War of 1812, which led to a surprisingly influential sponsorship of his application to be a cadet at the military academy. After the quick and disorganized studies that characterized those years of the academy, he served briefly in the artillery and then the corps of engineers before resigning to be a teacher at age eighteen. Closer in age to the cadets than his colleagues, Davies was a popular figure in the West Point community: "Of the middle size, with a bright, intelligent face, characterized by projecting upper teeth, which procured for him the name of 'Tush' among the cadets, his whole figure was the embodiment of nervous energy and unyielding will. His fearless activity at a fire which happened in a room in the South Barracks, in 1819, added the name of 'Rush on' to the other. He was a kindly natured man, too; and the patient perseverance that he devoted to the instruction of his class was not the least remarkable feature of his character."[25] Nicknamed "Rush on Tush," Davies was already a popular teacher at West Point, posed to use that good feeling to reform American math education more broadly.

Davies's West Point teaching proved most distinctive in expanding the geometry lessons that appeared at the beginning of Crozet's classes. For Latrobe, "It was with Professor Davies," not Professor Crozet, "that I began the study of descriptive geometry, for which no books in English had then been published. He had no assistance beyond the blackboard and his own intimate knowledge of the subject and faculty of oral explanation."[26] A young American with a modest education and easy personality, Davies was the ideal popularizer of Crozet's innovations. The blackboard, according to Latrobe, already had various associations: descriptive geometry, languages beyond English, knowledge, and talent at math communication—all of which Professor Davies embodied.

From engineering to geometry and beyond, the blackboard began to appear in various classes at West Point, gradually developing expectations for student performance. First in engineering and math, then in other fields, professors gave lectures at the board one day, expecting that students would take down detailed notes. Then students gave lectures at the board the next day, presenting it back to the class and their instructors once again.[27] Noticing such uses of blackboard for chemistry, scientist Joseph Henry noted in his journal, "Indeed it appears to be one of the principles of teaching in this institution that every thing as far as practical should be demonstrated on the black board."[28] What was distinctive, for Henry, was not just the presence of the pedagogical technology; it was that "the students" had to make these demonstrations.

Such expectations spread far beyond the military academy through Davies's exceedingly popular math textbooks. "It is to Professor Davies that I have always attributed in a great measure my subsequent successes at West Point," Latrobe concluded. "A much more enduring tribute is that awarded by the countless beneficiaries, the colleges, schools and individuals who have profited by his numerous publications in connection with mathematical science."[29]

Through these books, Davies addressed thousands, perhaps millions, and advocated that the nation's math classes adopt military models at a time when military academy pedagogy became synonymous with students' use of the blackboard.

Davies subtly and holistically recommended West Point math through his frequent references to French precedents and engineering applications. His first book, the 1826 *Elements of Descriptive Geometry*, encouraged the adoption of French methods in order to move beyond the synthetic proofs of Euclid to considerations of the geometrical constructions that could be drawn on spheres and "warped surfaces." Particularly in its applications, *Descriptive Geometry* had an expressly practical bent. Davies mentioned in his preface that descriptive geometry had an "intimate connexion [*sic*] with Civil Engineering and Architecture, and the facilities which it affords in all graphic operations, render its acquisition desirable to those who devote themselves to these pursuits."[30] As at the military academy, Davies's book taught the properties of geometric figures in order to motivate the drawing of three-dimensional objects in mathematically accurate two-dimensional maps. Like the introductory units of Crozet's classes, it led directly to methods in surveying and engineering. In order to garner a larger audience, however, Davies replaced the aim of training an "engineer officer" or military engineer with that of developing the "civil engineer" and architect. Still, the combination of French methods and applications communicated the precedent of the military academy.

Like the blackboard, the textbook series from Davies became emblematic of the ways math instruction could change to suit an engineering education. Previous textbook series had culminated in repetitive, idealized lessons in measurement, surveying, and navigation, and they seldom encouraged practical activities. Meanwhile, Davies reframed the entire course of math around the idea that it could prove professionally useful in direct ways. Davies's experience at the military academy exposed him to an educational model structured directly around students' postgraduation careers in the army and, at his time, the navy. The sense of utility and preparation inspired new questions about the traditional college-level subjects: How would geometry help a soldier build a better bridge? How would trigonometry help a seaman steer? How would calculus help a marksman tell where his shot will land? The curricular system from Davies and the military academy became increasingly popular in industrializing America. Davies's textbooks appeared throughout American higher education, and their accompanying terminology and assumptions still inform classroom practices. His encouragement of the blackboard especially has marked math education for many years.[31]

From these origins at the U.S. Military Academy, the blackboard developed certain expectations, especially for student performance. From French education, especially in military-engineering training, its uses expanded, though it still retained a focus on student-centered career preparation. Making possible the consideration of different kinds of (usually French) engineering and then mathematics, the blackboard likewise meant advanced education. Still, from *Arithmetic Made Easy* to *Davies's Series*, the blackboard was a tool of repetition: a place to repeat (accurately) the oral and written demonstrations of teachers and textbooks. At the military academy, as Phillips has argued, such student expectations allowed the demonstration of the cool mind of an officer, shown in his steady hand with the chalk and confident speech.[32] Beyond the military academy, students' facility with repetition translated to industrial careers.

## The Blackboard Elsewhere: At Trinity

Beyond the military academy, other colleges, scientific schools, institutes, and academies needed a different rationale for the blackboard's uses. After all, they dealt with students instead of cadets. Broadly, the pedagogical rhetoric that emerged in the middle of the nineteenth century combined elements of Yale's focus on liberal education with the U.S. Military Academy's orientation toward engineering applications and careers. In other words, colleges throughout the country started experimenting with curricula that ostensibly combined elements of liberal and technical education. These hybrid courses did promise to train some engineers, but they also (explicitly or implicitly) acknowledged the goal of providing leaders to direct the technological work of others. Even if the students did go on to work in factories, in mines, or in railways, they would not spend much (if any) of their careers on the floor, beneath the ground, or on the ties. They would instead direct the work pursued in these new industrial institutions from myriad offices, as historian Terry Reynolds notes, where they would pursue the vast project of determining how others should live and work so that the whole machine would run smoothly.[33] The expectations for public speaking associated with the blackboard, in the sense of calm oral reports for reviewing previously known information, found support in these emerging ideals of technological leadership.

Davies's career mirrored these changes in American education. He did not stay long at the military academy, and his subsequent appointments involved certain anxious arguments about how his kind of math could be useful. Over the course of nearly fifty years and at least four colleges, Davies kept trying to use French exemplars to reform the typical mathematical subjects: arithmetic, algebra, geometry, trigonometry, surveying, navigation, technical drawing, and calculus. However, after his *Elements of Descriptive Geometry*, Davies's books were not so explicit about their purposes for engineering education despite the fact that students were required to read them in nearly all the mathematics classes taught at every engineering school in the United

States. His own classrooms remained his best testing ground for his methods, but they looked very different after he left the military academy. After West Point, his next appointment took him to a traditional college of liberal studies, Washington College (now Trinity College), in Hartford, Connecticut, where he wrote the textbooks for analytical geometry, surveying, calculus, and arithmetic. Though Trinity's trustees had been experimenting with having a more scientific faculty, they remained committed to the common professional tracks of theology, law, and medicine.[34] Davies began to suppress the explicit arguments about civil and military engineering for general claims about the usefulness of math, especially math in front of a board.

An example of Davies's change proves to be his *Elements of Analytical Geometry*. He wrote it for a new sequence of required classes at Trinity College: in groups of about twenty each, the freshmen had to complete algebra and plane geometry, and the sophomores took solid and spherical geometry, trigonometry, surveying, and analytical geometry. (Calculus remained optional at the time.) Despite these institutional roots, Davies's *Analytical Geometry* began with sections outlining the importance of the military academy. He dedicated the book to Superintendent Thayer, who recently had resigned, and he explained that it popularized the French methods of geometry—and the blackboard—so prized under his leadership. Unlike his *Descriptive Geometry*, however, Davies could not be so explicit about the importance of geometry for engineering careers. After all, engineering was not even offered at Trinity. Davies instead ended with a disclaimer that introduced his thoughts about utility. He explained, "No attempt has been made to depart from clear and satisfactory methods adopted by others, merely for the purpose of seeming to be original." Instead, "It has been the intention to furnish a useful text-book."[35] For Davies, his preface introduced a "useful" book about a "useful" subject, stated vaguely and generally. But what were its uses?

His book's actual use was in math funerals (as in chapter 2) despite any theoretical considerations. At Trinity College, sophomores celebrated the completion of Davies's new analytical geometry requirement by creating a Latin ceremony with his textbook at its center. Their Concrematio Conicorum (Burning of Conics) began with an open procession with family, local friends, and freshmen in nightshirts. Bringing Davies's *Analytical Geometry* down to the riverbank, the sophomores placed the book on a funeral pyre, said incantations, played music, and sang songs. Throughout, they told geometry jokes. Then they burned the textbook, placing it on a boat for a flaming funeral or burying the ashes. In short, they committed the last ashes of their mathematics textbook to the land or sea in a move reminiscent of ancient funeral practices.[36] They likely borrowed the components of the ritual from Yale, Trinity's institutional rival, where the Burial of Euclid had already connected math to ancient Greek. Recall that such math funerals allowed a chance for actual students to practice their math communication. At Trinity, they practiced public speaking in Latin.

Latin, as opposed to Greek, signified a newly scientific curriculum. Even the name Concrematio Conicorum reflected Latin's new importance not for the study of ancient Rome but for careers in science and industry. At Yale and elsewhere, Greek had been questioned for its usefulness, but Latin remained necessary for its terminological applications, as historian Caroline Winterer has argued, especially for giving students the tools to decode taxonomies in botany, medicine, natural history, and even chemistry.[37] The Trinity sophomores, in drawing attention to the Latin roots of their ceremony, provided a system to explain their reliance not merely on ancient funeral practices but also on new scientific frameworks. After all, the textbook they burned presented the ancient study of conic sections through equations and formulas and their ends in engineering and architecture.[38] Latin wed the sciences in Concrematio Conicorum.

The name of Trinity's ritual furthermore suggested the changes that came from burning a textbook not from some long-dead mathematician but from a living, present author. Yale students mocked the figure of Euclid, calling attention to the connection between his book and body (*corpus*) throughout their mock burials. At Trinity, the textbook's author did not have a place. Trinity sophomores, after all, would not want to make a similar connection between Professor Davies's textbook and his body when he was in reach of the flames. In their title, therefore, they specified that "conics" would be burned but did not identify whose book. Even after 1841, when Davies left his teaching post to spend more time writing, the Trinity sophomores continued to be silent on the point of the identity of the burning textbook's author. In such a move, the image of the mathematician-author faded into the background and that of the subject became visible. The funeral service, Trinity students maintained, was for conics and not its interpreter.

In short, the Trinity ceremony exposed what happened when teachers brought West Point methods to colleges that did not have officer training or engineering. Trinity sophomores, after all, recited in Latin while holding aloft an engineering-oriented, French-inflected textbook. Their college had gone so far as to require classes in conic sections, yet it did not relax many of the requirements for ancient languages as the military academy had. When Trinity students performed their college identities in a public display, they incorporated both mathematical applications and Latin. Davies's *Analytical Geometry* proved literally central to their celebrations, except their ancient allusions communicated the ways that their college still respected and replicated older schemes of liberal education. Even when Trinity faculty maintained a careful differentiation from Yale's through including more scientific offerings, they did not go so far as to reject the humanities in favor of officer training or even engineering. As the students' reactions demonstrated, the sciences gained prominence in their educations, but they by no means took over.

Math funerals reflected expectations for the blackboard even when the blackboard was not used as a prop. After all, math funerals were occasions for mocking the terminology of required math classes, making a show of knowing something about college life. Through them, actual students repeated their teachers' blackboard lectures back to them. Practicing math communication, though not at a board, these students followed the West Point model for reviewing math information after it was learned. At Trinity, students used Davies's expectations for blackboard behavior to speak back to his own classes and books.

As Davies continued to travel from one academic appointment to another, he brought the blackboard (and its rules) with him, and he also brought his increasing uncertainty about the uses of West Point mathematics. Between his appointments at the University of the City of New York (now New York University) and Columbia, he decided to publish an unlikely addition to his college series: *The Logic and Utility of Mathematics*. For Davies, it served to answer the questions of his vast readership. He began the book by indicating its roots in his reflections about how the military academy taught math differently from other colleges, and he quickly characterized their "French" style of instruction (with the blackboard) versus the "English" style at other American schools. In order to explore these differences, he introduced the terminology of logic (induction, syllogisms, etc.) before showing how logic undergirded the construction of arithmetic, geometry, algebra, and calculus. Finally, after three hundred pages about such topics, he returned to the question of the usefulness of mathematics.[39]

The "utility of mathematics" in Davies's title ultimately connected to the development of rules, implicitly recommending the "nobility" of math repeated at the blackboard. What were math's uses? The simplest answer was that it provided "intellectual training and culture." Parroting the adherents of mental discipline, the ones who supported chapter 1's recitations, Davies quickly summarized the ways math could strengthen the mind. He also built on the broader popularity of French styles by indicating how "intellectual culture" came from explicitly French sources. Second, Davies stated that any field of math could be a way of "acquiring knowledge," particularly according to Francis Bacon and others in the "English" tradition. Last and most importantly, mathematics could provide "rules" that were "practically effective." Though expressing discomfort with the popular terminology of "practical" education, as Jeremiah Day had done a generation before, Davies still listed examples of subjects and their applications: arithmetic in commerce and building and more advanced math in factories, steamships, surveying, railway engineering, and reservoir construction. In describing these applications, Davies also equated mathematics with industrial leadership. Math oversaw factory work like a "governor machine." It acted as a steamship captain. It motivated the surveyors and engineers who planned the technological fascinations of the age, including (Davies's favorite) the Croton reservoir serving New York City. In short, more Americans needed to recognize math's inherent value—as he said, "nobility"—for processes of industrialization and development.[40] Coming from his experience at the military academy, Davies's view of mathematics stemmed from the training of "noble" officer-engineers at blackboards. For Davies, mathematics remained the ideal preparation for industrial leaders who would advance the aims of the United States even if those aims were sewage disposal or clean drinking water. The blackboard epitomized the French, scientific, military origins of the system of math as well as its new possibilities.

Beyond Charles Davies, advocates of the blackboard had increasingly indirect connections to the military academy. Harvard professor John Farrar began to use blackboards in his geometry classes in the mid-1820s, and he built assumptions about their use into his own textbook series. When he published *Elements of Geometry* in 1823, he asked that the plates be published separately so that students could consult them (and not the book) when presenting at a blackboard. Bowdoin professor William Smyth advocated for the use of the blackboard, too, in part because his algebra lectures at the board led to his campus popularity and ultimate promotion in 1825.[41] A technology of classroom communication, the blackboard clearly involved expectations for both teachers and students, all of whom were supposed to act calmly and powerfully in reviewing predetermined information.

## Clarifying Rules for Math and Beyond

Not only at Trinity but also at Yale, the blackboard's introduction coincided with a linking of ancient and mathematical subjects. While Trinity's students expressed the marriage through their Latin mathematical funerals, Yale faculty asserted the connections between mathematics and ancient Greek. Beyond West Point, they argued, the expectations for speaking like a calm officer-engineer translated to the liberal leadership of "mental discipline." Popularized in "The Yale Report of 1828," these statements codified and naturalized the new expectations for math communication while also extending their influence far beyond the math classroom.

The report, in short, justified college education through the case of mathematics. The year before, Connecticut state senator Noyes Darling had asked if Yale's faculty would approve the substitution of modern languages for ancient ones, since he thought them more suitable for the nation's emerging market economy. Later, by way of clarification, Yale alumnus Thomas Smith Grimké said that an education founded on ancient Greek and Latin was "totally useless to the great majority," not only the "working classes" but also the "man of business."[42] Why not teach the modern languages instead? So many prominent alumni repeated the same question that Yale's corporation and faculty appointed five men to consider the problem. Two ministers, the Connecticut governor, and Darling formed the external branch of the committee with classicist James Luce Kingsley and President Day representing the college. Quickly, the committee decided to move beyond the mere consideration of the classics to a justification of Yale's curriculum in general. Both laying out a detailed defense of the traditional liberal arts and locating Greek and Latin as central to that system, the report is remembered for its vindication of the academic status quo, as historian Caroline Winterer states.[43] Despite such attention, the report's historiography has not emphasized the role of math communication.

Math's essential role, in fact, remained subtle. As noted in chapter 1, Day's *Algebra* textbook already assumed the universal applicability of mathematics. The report had the same assumptions, and the values associated with mathematics justified the preservation of ancient languages. Since math's importance was assured, the committee argued by analogy for the security of ancient Greek and Latin. Even more than Day's *Algebra*, the report naturalized the perceived values of math because math was not the actual topic under consideration. Fundamentally, the report built on the rhetoric of mental discipline, extending it from geometry to the ancient languages, which not only explicitly argued for the importance of Greek and Latin but also implicitly justified geometry. In a lasting way, the report truly established and popularized expectations for math communication within a college context.

Mental discipline, at the time, remained a staple of Yale education, particularly in its connections with mathematics. As mentioned in chapter 1, the Yale faculty and their contemporaries looked to Locke's *Of the Conduct of Understanding* for the foundations of a system of mental exercise. "We are born with faculties and powers capable almost of anything, such at least as would carry us further than can be easily imagined," Locke wrote, "but it is only the exercise of those powers which gives us ability and skill in anything and leads us towards perfection." Such mental exercises, in Locke's system, therefore allowed unlimited (or nearly unlimited) improvement of the "faculties and powers" that make up the mind at birth. So what did Locke recommend? "As in the body," he continued, "so it is in the mind; practice makes it what it is, and most even of those excellences which are looked on as natural endowments will be found, when examined into more narrowly, to be the product of exercise and to be raised to that pitch only by repeated actions."[44] Locke therefore asserted that practice and repetition could allow mental muscles to appear strong, even "natural." Locke advised, in modern terms, that practice makes perfect, and he found the perfect calisthenics in mathematics.

At the time, the Yale report argued by analogy from the assumed power of math in order to indicate the skills that would be gained through the study of ancient languages. As a first step, the committee asserted that the values of judgment, argument, and persuasion were inextricably tied to math studies. Specifically, geometry requirements made sure that the student's intellect became attuned to "the discovery of truth and detection of error." Repeating Locke, they assumed that a student repeatedly working with Euclidean proofs would come to understand (perhaps not right away) that such ideals could be applied to nonmathematical arguments. Someone who developed a taste for geometry would be able to choose which arguments most resembled abstract proofs, understood to be the best supported. In later life, he could use these skills to be a better judge of any assertions and to be a more logical, more persuasive communicator. "Whether his own station in life is public or private," they summarized, "whether he engages in a professional career, or is called upon to discharge the duties of a magistrate, the occasions for employing his knowledge are innumerable."[45] In other words, skills gained through math not only prepared a student for the concerns of a career; they also strengthened him for his leading role in his community and "private" life. The report claimed that Greek and Latin could train him too.

Closing their argument, the authors of the Yale report explicitly demonstrated that the same mental discipline could come from the classics as well as mathematics. Latin and Greek also "discipline[d] the mind" through the repetitive actions of declensions, conjugations, and even translations. They also taught argumentation through the exemplars of Plato, Aristotle, Cicero, and Caesar. Judgment and persuasion formed key components of these classes, and the ancient languages had an even clearer connection to the university's professional schools (i.e., law, medicine, and theology), all of which relied on classical terminology but rarely used geometric proofs. Therefore, the report concluded, since "the usefulness of mathematics is in general admitted," so too should the usefulness of Greek and Latin.[46]

Reinforcing the point, they implicitly reasserted the value of math by presenting their arguments in a form reminiscent of a syllogistic proof. They began from the well-regarded virtues of math, stated explicitly from what they believed to be tacit assumptions held by everyone "in general":

(A) Mental discipline is useful in education.

They then demonstrated that the same virtues also came from the study of ancient Greek and Latin, concluding that

(B) Studying ancient languages is equivalent to mental discipline.

Which meant, therefore, that

(C) Studying ancient languages is useful in education.
QED: quod erat demonstrandum.

Alluding to a common proof form, the committee explicitly argued against proposals from the likes of Darling, and they implicitly reasserted the importance of math, not only for their students' arguments but also for theirs. Math communication, for them, meant not only a calm voice and a measured hand; it also meant logical argumentation that could be reducible to a proof. Such skills developed through practice enforced by the student and by the college as a whole.

The publication of the report not only popularized such expectations for math communication but also linked them to Yale. For the local community, Hezekiah Howe printed a version called *Reports on the Course of Instruction in Yale College: By a Committee of the Corporation, and the Academical Faculty*. Howe's shop, the first issuers of Day's *Algebra*, had expanded to include a wide range of books purportedly for large markets extending throughout New England: cookbooks, theological treatises, naval and astronomical almanacs, sermons, children's stories, and travel narratives. Howe's printing allowed the committee to begin disseminating their arguments to Yale alumni and friends. Senator Edward Everett and Mayor Josiah Quincy—prominent Bostonians who would both become presidents of Harvard—received copies. So did the brothers Ebenezer Porter, the president of the Andover Theological Seminary, and Thomas Porter, a Hartford attorney. Within four months, it became necessary to authorize another edition, which appeared in 1830. For a little while, it was the talk of the law office, the school, and the political hall. The report served to reinforce bonds between Yale and its affiliates throughout New England, as noted by many Yale chroniclers and educational historians.[47] The report asserted that Yale had created a model for others to follow: explicitly in "instruction" and implicitly in and through mathematics.

Day's report further promoted Yale's mathematical expectations through its publication in the *American Journal of Arts and Sciences*. The *American Journal* was not only the leading scholarly periodical in the nation; it also brought academic recognition to Yale through its well-known link to the college's first science professor, Benjamin Silliman. Initially financed and edited by Silliman, the journal had even acquired the popular nickname "Silliman's journal."[48] Another copy of the report appeared there as "Original Papers in Relation to a Course of Liberal Education," listed only with the authors Jeremiah Day and James Luce Kingsley. Occluding the rest of the committee except for the two representatives of the Yale faculty, the *Arts and Sciences* version further underscored the association of the text with Yale itself. With a subscription list in the hundreds, Silliman's journal sent the report throughout the country: not just to Connecticut and Boston but also to Philadelphia, western Pennsylvania, and Kentucky. The article and pamphlet quickly gained the moniker "Yale Report," and it asserted the intellectual primacy of Yale for decades to come.[49] Still, though admittedly about Yale, the publications made the curriculum less specific, more like the kind of curriculum that could be adopted anywhere that practiced a "Course of Liberal Education." Its circulation reinforced the rules of college life, subtly extending Yale's model and extending outward from expectations for math behavior.

On the eve of the Conic Sections Rebellion of 1830, therefore, Yale affiliates reified and popularized rules for "natural," "discipline[d]" math communication with rationales "innumerable" and "in general admitted"—all incompatible with expressions of stage fright at the blackboard. Beyond the sense of calm communication promoted through the ideals of an officer-engineer with a steady hand and measured voice, Yale educators asserted mental discipline's valuation of mathematics. In their view, repeated problems in front of a board should lead to more logical communication: strengthening students' mental faculties, argumentation, judgment, and persuasion. It should encourage their development into leaders. It should not encourage students to act out. When students disobeyed in the Conic Sections Rebellion of 1830, the Yale faculty responded by saying that such a thing was impossible within math, within their college, and within the whole of American education. The problem was with the students—who were henceforth kept out of all academia—not the expectations for using new classroom technology.

## Debating Stage Fright at Yale

The military academy rules for the blackboard, even when changed at other schools and colleges, did encourage stage fright in students. After all, as Joseph Henry remembered, West Point's distinctive "principles of teaching" involved student demonstrations, reviewing the previous day's material in front of the board. At the time, some students felt that the form of public speaking placed unfair expectations on them, ones that previous generations of students did not have to encounter. Coupled with textbooks like Davies's, Farrar's, and Smyth's, along with *Arithmetic Made Easy* (a product of the Anglo-American movement to teach "poor" children), such systems did make some unflattering assumptions about students: about their socioeconomic background and/or about their career prospects. Because of such associations, there was a student outcry when Yale faculty introduced the blackboard. From specific cases of stage fright in math class, the conic sections rebellions involved bigger arguments about students' power with respect to the new classroom technology and its expectations for their communication.

President Day publicly proclaimed the importance of obedient, "disciplined" math students, and his stance paradoxically led to a surge in student rebellions. When Yale students matriculated at the college after independence, they agreed to various rules, including the following: "on [my] faith and honor, [I will] obey all the laws and regulations of the College; and particularly, that [I] will faithfully avoid *all combinations* to resist the authority of the Faculty."[50] Such a pledge was necessary because riots and rebellions were common features of life in the young republic.[51] The Yale faculty understandably did not want a riot on their hands, but it is unclear how much the matriculation promise gave students ideas. By 1829, Yale students had already "entered into combinations" at least twice before, according to Yale historians.[52] When they rebelled again, in the Conic Sections Rebellion of 1830, Day's reactions and students' petitions conflated geometry requirements with discipline of the student body, as we saw in chapter 1. These documents also showed how new rules for communicating mathematically followed the introduction of the blackboard.

The 1830 Conic Sections Rebellion began with the new expectations for reading with the blackboard. After the "Yale Report" established the college's importance, the students had new requirements placed on their time, including a new, sophomore-level conic sections class. Rather than complaining about the requirement outright, the sophomores protested the new policy. The blackboard encouraged a new publishing practice: that figures could be printed separately so that students could review from them (and not their books) in front of the board. At Yale, the practice became a formal policy: in conic sections, blackboard demonstrations had to occur from diagrams. But the students wanted to use the printed description in their textbook as a memory aid; maybe they were embarrassed to recite without it. They submitted their first petition on Wednesday, July 28, 1830, and when they did not get a response by the time of their next class (Thursday morning), they decided to act. The tutor for the 11:00 a.m. class called on individual students to participate as usual. But eight or nine refused. Finding the students insufficiently obedient, the tutor canceled class. The policy of boardwork remained.

Students and faculty, on Thursday, created quantitative studies to demonstrate the source of the worry, apparently measuring the extra work the sophomores had to do. Even before the faculty could meet to deal with the first petition, they received a second one, suggesting that the "lessons in conic sections might be shortened." It claimed that students had already spent more time in math than previous generations: Algebra was "one-half more difficult" than previous classes. Trigonometry was incomparably more difficult. In short, "none of our lessons have been easier," they argued, and "some have been longer and harder than usual."[53] Surprisingly, the faculty did not reject these demands outright. Accepting mental disciplinarians' arguments that math was important for its connections to nonmathematical argumentation, the faculty checked the figures. Preparing charts of the work assigned in that year's conic sections class and comparing it to similar measures from a smattering of previous years, they found that Yale math had not changed significantly, at least not enough to warrant special favors to the current sophomores. "The Faculty have not found," they concluded, "that the length of the lessons given to this class in conic sections, is greater than the average length of those given to the five or six preceding classes who have recited the same work; or that greater burdens have been laid upon this class, than upon others." Though the students tried to quantify "difficult[y]" in front of a blackboard, the faculty responded by checking content and creating a measure of "burdens."

Discipline remained the key issue in these blackboard debates, but student disagreements emerged about the importance of obedience versus loyalty. Should they follow the "laws of authority" of the faculty, or should they follow each other? A representative of the class delivered a third petition to the faculty. It noted their disregard for university rules but asked for equal punishments for the entire class body. "We the undersigned, members of the Sophomore class," it began, "hereby declare, that if we had been called upon by the Tutors, yesterday morning, we should have recited from the book and not from the figure; and that we will render ourselves obnoxious to the same punishment, with the individuals who did recite, by transgressing every college law, until it is adjudged."[54] Threatening future action, the petition noted the problem stemmed from general university rules. While the suggestion of menace did alarm certain faculty members, the signatures were alphabetized and seemingly written in the same hand. In other words, the petition was apparently about broad-based "punishment" for the whole class, but its presentation suggested the work of only a few spokesmen trying to represent everyone. Meanwhile, twenty-one students, half of those who supposedly signed the petition, submitted promises to Day and the rest of the faculty: "We the undersigned do hereby declare our readiness to recite Conic Sections in the manner prescribed by our instructor."[55] Not specifying "the manner," these oaths implicitly represented a promise to follow any university prescriptions. Explicitly, they were declarations of obedience: obedience to the math instructors and obedience to the rules of blackboard use.

Still, the subsequent actions of certain anonymous students intentionally confused the faculty about who remained obedient and who did not. A fourth petition tried to address the strange format of the third petition, saying that the anonymous "committee" of authors had received permission from the rest of the class to copy out their names. It additionally argued that the sophomores could be both obedient to the faculty and loyal to each other at the same time. Meanwhile, on Saturday morning, the faculty decided to dismiss three students who they thought were the principal leaders: "Whereas Edward

Reed [and John Steiner and Elijah H. Hubbard] of the Sophomore Class has taken an active part in enciting [*sic*] opposition to the college government; and has promoted and entered into a combination to resist the authority and laws of the college; it is therefore determined that the said Reed [and Steiner, and Hubbard] be, and he is hereby dismissed."[56] Noting the specific problem of "a combination" resisting their "government . . . authority . . . and laws," the faculty no longer engaged in quantitative arguments about "burdens" but instead decided to act on the matter of obedience.

Reflecting the origins of the conflict in communication rules, some unnamed students then clarified how the petitions should be read. Before a representative of the faculty could deliver the sentence to Reed, Steiner, and Hubbard, a fifth petition appeared. Still worrying about the fate of the nine who did not recite, it noted their actions represented the wishes of the whole class. In explicit language, it talked about how the sophomores therefore should not be treated as individuals. Indicating "perfect equality with them to participate in their destiny whether that destiny be continuance in the College, dismission [*sic*], or expulsion," the petition noted the importance of judging the forty-two distinct signatories as a uniform whole.[57] Explicitly, it presented guidelines for judging these students' written communication, and it asked the faculty to abide by their rules.

But they went too far. In punishment, the faculty focused on class-level disobedience, and they acted accordingly. Not only did they dismiss three students, but they also wrote letters to the parents and guardians of the forty-two sophomores that had signed any petition. They punished all of them by asking them to leave the college grounds as soon as their guardians arrived. The students, hoping to protest the ruling, appeared in large clusters before the faculty on Monday and Tuesday, but their actions merely indicated that they continued to disobey college rules through these new groupings, these new "combinations." Though first about "burdens" in front of the blackboard, the complaints now represented the willful, sustained neglect of Yale laws.

Day's subsequent correspondence emphasized his expectations for obedience, in math class and beyond. Just weeks after the expulsion of most of the sophomore class, Day received letters from dozens of fathers asking for advice about their sons:

The unexpected return of my Son from new [*sic*] Haven was very surprising & under the circumstances very painful . . . I write to request information, whether . . . the Faculty will consent to determine him in good standing so that he can join some other college.

oblige me, Sir, by answering this letter . . . in order that I may fully determine the course that I am to propose.

I take the liberty of expressing a like, that . . . the members of the class . . . pursue their studies, in the regular manner.

inform me what you may deem proper for my Son to do.[58]

Such requests reaffirmed Yale's commitment to act *in loco parentis* (in place of the parents), particularly in matters of discipline and academic life. These fathers asked for advice from Day about how they might properly educate and raise their sons without his continued guidance. Day responded by popularizing his version of the story.

In correspondence with presidents and faculty of other colleges, Day asserted Yale's exemplar status as disciplinarian. Josiah Quincy, the new president of Harvard and a notable recipient of the "Yale Report," proposed that Yale's decision held at other colleges as well: "I have had several applications to know whether the malcontents could be admitted here. Of course, I had but one answer: 'When they brought a certificate that they had taken up their connections [with Yale] in good standing and to your satisfaction, their case might be considered; without that never.'"[59] Yale's laws, for Quincy, stood in for the laws of the whole "community" of educators, that Yale's decision about their "malcontents" should hold throughout American colleges. A college president in Coxsackie, New York, wrote that he agreed, though for a different reason: the "malcontents" still had connections with Yale, albeit tainted ones, and they should not be connected with two colleges at the same time. The following summer, when the president of Middlebury tried to disagree, Day noted his stance. "I would observe," he said, "the maintaining of proper discipline, to be a common interest of the colleges, and not a separate concern of our own."[60] In other words, any college should respect the punishment delivered at Yale. In these exchanges, Day—and, when he was absent, Benjamin Silliman—successfully established Yale as setting the standard for discipline throughout American academia, from obedience at the blackboard to obedience throughout college life.

From the summer of 1830 to the summer of 1831, Day and Silliman reasserted Yale's control by retelling their story about how the students acted inappropriately in front of the blackboard and beyond: circulating letters and ordering the printing of pamphlets about the incident. Presidents, professors, and parents, they claimed, should know about the events at Yale: how they began with a simple neglect of the new blackboard rules and became a full-blown revolt. The college never did make a formal declaration about the students' subsequent status, but because of Day's efforts, none of the forty-two could return to Yale and almost none could go to any other college again. For Yale historians, the class of 1832 still remains a blip in the college records, showing a time when only about half of the class made it to graduation.[61] For the students and their families, the Conic Sections Rebellion ruined college plans and career prospects because of the new expectations of a new classroom technology.

Day capitalized on the event to argue against the students' sense of stage fright too. In Day's version of the story, experiences of anxiety were not possible in the math classrooms he oversaw. According to Day, his students presented themselves as men of "ordinary intellect" and objected to the new requirement of conic sections and the new blackboard rules. In short, Day thought that the students committed two main errors. First, and fundamentally, the sophomores used their stage fright as an excuse to ignore college rules. They ignored the stipulation that they should not "enter into combinations," and their collective actions resulted in what the pamphlet euphemistically called "the recent proceedings of the sophomore class."[62] Second, they ignored the reasons they learned math in the first place. Neglecting Day's frequent invocations of mental discipline, they did not understand the ways that geometry already combined exercise and the search for mental perfection. In other words, geometry was supposed to transform students' "ordinary intellects" into something extraordinary by the time they reached conic sections. The students should not have experienced stage fright in the first place; for Day, it was simply illogical.

From the standpoint of today's medicalized anxiety, Day's counterreaction to the blackboard is itself perplexing. Anxiety, even fright, comes with certain physical signs, such as rapid heartbeat and labored breathing. Such experiences are considered pathological—or at the very least, uncomfortable—because they cannot be fully controlled. The Conic Sections Rebellion began with experiences that the students perhaps could not control, though they tried to quantify them. Unfortunately, for them, the new rules for blackboard communication expected a high level of control over their bodies and minds. In their worry, these students already began to seem "inappropriate," and perceptions of their behavior quickly spiraled—from disobedience to rules for new classroom tech, to disobedience to their teacher, their whole faculty, their whole college, and even the whole of American academia.

## Math on Stage

Notably, the Yale students could also begin to experience stage fright in boardwork because they were beginning to consider math as occurring on a stage. The use of mounted boards created a place to draw attention, a place to gather, as *Arithmetic Made Easy* put it.[63] When used often enough, the arrangement emphasized a certain layout of the classroom: chairs arranged so that they faced the blackboard. When Davies, Farrar, Smyth, Day, and many others encouraged students to demonstrate, the students were suddenly on display, too, at the locus of everyone's gaze. A generation later, just into the second period of American blackboard use, the 1858 lithograph of the Burial of Euclid showed how much students at Yale understood these expectations for performing math, representing math work as theatrical and math on stage.

As noted in chapter 2, the Burials of Euclid started around the same time as the two conic sections rebellions, though they were not entirely earnest or rebellious. Yale's math funerals similarly responded to expectations for obedient math communication. Still, compared to the Conic Sections Rebellion, costumes and songs replaced the petitions, and Burial "mourners" did not submit statements about what made math "difficult." Instead, students seemed to revel in the fact that they had superior geometry educations, better than previous generations, better than the freshmen, and better than the New Haven residents. In the Burials of Euclid, the sophomores both complained about new math requirements/expectations and also celebrated their accomplishments. The Burials of Euclid, too, placed students on display, yet unlike with the blackboard, they were the ones who chose to perform.

The Burials of Euclid performed class pride by mixing mathematical and Greek references, reminiscent of the "Yale Report." The 1856 *A Collection of College Words and Customs* included a parody of "Auld Lang Syne" that culminated in ancient Greek words of mourning:

Though here we now his *corpus* burn,
And flames about him roar,
The future Fresh shall say, that he's
"Not dead, but gone before":

We close around the dusky bier,
And pall of sable hue,
And silently we drop the tear;
"[feu feu oi moi, feu feu]."[64]

About both Euclid ("he") and the class ("we"), the song began with the notions of math class and continued to the previous verses about Euclid's funeral. Student costumes reiterated these allusions. Some took on the characters of Euclid's family: his stepmother "Parent Hesis" (parentheses), his aunt "Aunty Cedent" (antecedent), his daughters "Polly Gon" (polygon), "Anna Lytics" (analytics), and "Cora Leary" (corollary), and his son "Geo. Metry" (geometry). Others acted as the tutor "Marcus Low" (mark us low) or the professor nicknamed "Pilot of Asses," an adaption of the popular nickname for the first book of Euclid. Vaguely classical, the costumes were eminently mathematical. *College Words and Customs*, underscoring the point, included not only the description from *Sketches of Yale* (quoted in chapter 2) but also an article from the *New York Tribune*. Celebrations, it said, took place in the local Masonic lodge, partially for the classical-mathematical ambiance and partially for the theatrical layout.[65] On the Masons' stage, the Burial of Euclid characters could perform for their audience of "future Fresh."

In 1858, a Burial of Euclid lithograph represented a combination of ancient and math references on/through a proscenium stage (see figure 4). From W. H. Davenport, a New York City artist who never finished a Yale degree, it was a piece of Emil Crisand's larger series about Yale college life, and its advertised audience consisted of students, alumni, and supporters of the college.[66] The lithograph encapsulated the amalgamation of classical and mathematical allusions inherent in the Burial of Euclid. In the image, the students occupied the same space as pegasi, figures from classical and contemporary allegory, and Zeus wielding what were labeled "tangents and cycloidal curves" instead of lightning bolts. A moonlit scene, it was dominated by smoke rising from barrels surrounding a coffin. The cheering crowd, dressed in various styles and colors, watched the conflagration. The smoke, rising above their heads to the right, then became part of a proscenium frame, first aiding angular devils in carrying a Jesus-like Euclid to the skies. The smoke—becoming a proscenium curtain—rose to Zeus's throne, where he averted his eyes from the scene, throwing his arm to the left column of students on hobby horses (made of textbooks). Meanwhile, two representations of students appeared at the bottom corners: the left representing sloth and the right representing the sickness/pallor that comes from overstudy. They sat above two insets, one showing a professorial figure with a book and the other showing a party, perhaps juxtaposing their educational illnesses with possible cures. The bottom of the proscenium stage identified the scene as "Yale—The Burial of Euclid." A crying crocodile split the text, separating the *YA* from *LE*, and its presence suggested the scene as parodic, not an actual funeral but a mocking one.[67] Davenport's lithograph therefore indicated the complexity of the event: combining allusions celebratory and riotous from ancient languages and from geometry. All were ultimately on a stage; all were ultimately theatrical.

Davenport's lithograph became the unofficial symbol for American student traditions. The lithograph's description appeared in print in the 1871 *Four Years at Yale*, an anonymous book that promised the "undergraduate," self-consciously unofficial point of view on Yale life. Like the earlier *Sketches of Yale*, it too contained sections about the college's history and its society system, student life, and curriculum. Unlike *Sketches of Yale*, it tried to follow students from freshmen through senior years. Explained as the "oldest distinctively student custom," the Burial of Euclid is featured in the section about the sophomore year.[68] Combined with lengthy quotes from *Sketches of Yale*, student stories, and a story from the *Yale Literary Magazine*, the lithograph came to represent "student custom" at Yale and beyond.

FIG. 4 "Burial of Euclid" (1858) lithograph by Emil Crisand, W. H. Davenport (del.). Personal collection.

What about American college life did it represent? In a certain sense, the Burial of Euclid reinforced the exclusive character of college education at the time. The entire burial practice was one large allusion to ancient practices, with orations, processional, funeral pyre, and in at least one case, ancient Greek words of mourning. The costumes, too, self-consciously echoed their understanding of ancient civilizations; *Four Years at Yale* called them "something like that of the old carnivals at Rome."[69] In a certain way, these layered allusions acted to keep others out. Though spectators were essential to their celebrations, the sophomores also incorporated symbols that communicated intense secrecy: the Masonic lodge, passwords, and even a "committee" acting as the event's security. At the time, when less than 1.5 percent of the American population attended college, the act of incorporating blatantly collegiate subjects—and performing a certain level of secrecy about them—reaffirmed the students' privilege.[70] In short, the Masonic stage not only telegraphed the theatricality of math at the time; it also placed a certain distance between students and their (anticipated) audiences.

The tropes of drama also helped indicate the complexity of the occasion for an audience never present: the faculty. While students did dress as tutors or professors and purposely processed through the streets where faculty lived, they did not construct the ritual as a protest. They printed small blue pamphlets to show the "Programme of Exercises," from overture to Latin ode, oration, poems, songs, procession, prayer, and dirge. They did not sign petitions or publish demands. While the symbols of the Conic Sections Rebellion suggested a riot of the American Republic, those associated with the Burial of Euclid did not communicate hatred or even displeasure. In fact, as in Davenport's lithograph, the features indicated mixed emotions, all through a proscenium frame.

The smoke, too, did theatrical work in the lithograph. According to its paths, the ceremony represented an elaborate transfer of knowledge, one similar to but at odds with Day's mental discipline. On the one hand, the sophomores, like Day, constructed mathematical knowledge as intertwined with classical allusions. The drawings, songs, incantations, and even passwords emphasized the interconnections between geometry and ancient languages. Even by referring to the ceremony as the "Burial of Euclid" and not, for instance, the "Burial of Day," they reaffirmed the importance of an ancient mathematician. On the other hand, the students were no pillars of the community. Though not openly protesting their academic requirements, they were also not following university rules, especially after an 1843 rule banned the practice. Like the sophomores in the center of the lithograph, they were negotiating a space between the studious though sickly and the strong though slovenly. While they appeared to acknowledge that some amount of a college education could give them power, they also performed a belief that excessive study sapped their strength. On the periphery of Yale, the sophomores performed interconnections between geometry, Greek, and cultural power. Placing math on stage not only allowed them to indicate all these associations but also allowed them the option, if pressed, to explain them away as just a (smoky) joke.

The blackboard did, in certain ways, make possible an explicitly performative sense of mathematics. As students and faculty debated the relevance of math's relationship to older studies, the blackboard reasserted math's values for society. It encouraged a system of lecture and review where everyone went to the front of the room, everyone supposedly acting calmly under pressure. In a certain sense, it is not surprising that some math students produced cultural artifacts that involved a stage and that some tried to claim (even quantify) stage fright in front of the board. Just before Davenport's lithograph and just before the heyday of math plays (in chapter 4), the blackboard's introduction already encouraged a redefinition of what it meant to be performing as a math student—not only in ways indebted to reading, textbooks, and idealized notions of studenthood. Making possible certain ways of speaking and excluding others, blackboards re-created learning math as about display. Building on the already redesigned space of the math classroom, now complete with boards and platforms, students at Vassar College and elsewhere went on to recraft student responses to math class, literally putting math communication on stage.

# 4

# How Math Communication
# Has Been Theatrical

• • • • • • • • • • • • • • • • • • • •

Davenport's imagined stage contained more than students. Ten characters appeared just behind the red Yale banner, variously reacting to the smoke and pyrotechnics. Eight of them, all stereotypes, faced the action from the sides: a Connecticut Indian with a sketched guide and three tradesmen (with various hats) on the left; a merchant (with hooked nose), a military man in uniform, and a policeman (with torch and bell) on the right; and an elderly man in academic and/or priestly robes in the center, turned entirely toward to the action. Meanwhile, another character, in sea monster costume, glanced back from the line of audience members, drawing attention to the crying crocodile—to the self-conscious humor of the occasion.[1] While the Burial of Euclid lithograph mainly contained representations of students, imagining them in the middle of burning, riding, preaching, and drinking, it clearly included an audience as well. With expressions variously entertained, annoyed, and occluded, the spectators served a clear purpose in these rehearsals for math communication. The students were showing off, and that action needed to have an audience.

Focusing on the role of exclusion in mathematical performances builds on recent scholarship that combines approaches from gender history and history of education. After the Civil War, more young women started attending school and encountering math at higher levels, to the point where their math success became literally threatening to their male peers and educators, as historians Kim Tolley, John Rury, William Reese, and Ellen Condliffe Lagemann have argued.[2] While school leaders established quotas to preserve boys' egos, male students increasingly separated themselves from their female peers, claiming that they were different from and better than the women, according to chroniclers of "co-educational" institutions.[3] As archaeologist Laurie Wilkie has noted, these rulings (from men) specified how women could seek education and employment, segregating them at a time when men and women socialized more than ever.[4] While the 1850s lithograph illustrated class-based and race-based exclusion from the mathematically inspired "stage," the next generation of students acted to separate women. Among the male-only spaces claimed at certain schools and colleges was the math classroom. In addition to sports fields, rallies, and fraternities, some performances of math communication developed to preserve men's sense of their right to their own space.

Such analysis similarly adds a historical perspective to the sociological work about the gendering of mathematical success. Cultural scholar Sara Hottinger has analyzed the ways that current mathematical textbooks at various levels constrain "our [American] cultural understanding of mathematics" to be one of white masculinity.[5] Heather Mendick and Valerie Walkerdine similarly have explained the attrition of high-achieving (British) women in mathematics by studying female students' difficulties in reconciling their math success with their identities as women.[6] A continued analysis of the place of gender in student performance should investigate connections to literal performance traditions.

Mathematically themed student plays of late nineteenth-century America acted as an extension of the informal performances of math textbook funerals (from chapter 2). Gender exclusion was often featured in these performance traditions of male college students. Young men at Columbia attempted to perform their collegiate knowledge before an audience of women at Rutgers Female College, and male students at Cornell showed off mathematically for the women of Wells College. Such performance traditions, I argue, developed certain characteristics to keep their female peers separate and silent.

Still, some women claimed their own opening night. Outsiders already closely observed female students at schools and colleges, in what was touted to be a vast "experiment" to see if they had the intellectual capabilities of men.[7] Already placed on display in the math classroom and far beyond, some chose to take to the stage, writing and acting in plays about mathematics. At Vassar, a self-consciously leading women's college, theatrical math importantly encouraged the acceptance of women as college students. Through the characters of math and its study, these plays convinced men at established institutions, and even the humor of these productions held a serious point. Male students saw themselves in the plays, subtly expanding the bounds of college community to include women. Moreover, because math was not merely peripheral in these productions, Vassar students proved themselves to be college students through their oral and written communication about math topics. For these students and far beyond, math communication has been not only performative but also literally theatrical.

## The Gilded Age of Math Funerals

After the Civil War, the rehearsals for math communication went on stage. The math funerals, already performative, became performances in their own right, and students at many colleges supported the show. The fighting of the Civil War, which mainly took place on the lands of the former Confederate States of America, meant that Southern colleges struggled to remain open, since many students had left for battle.[8] The Northeastern United States, however, entered the financial heyday that Mark Twain and Charles Dudley Warner parodied as the "Gilded Age," implying that the glitz of the era remained on the surface only.[9] The wartime support of the railway and factory created a class of fantastically wealthy businessmen who looked to institution-building as ways of spending their vast sums of money. Ezra Cornell, Matthew Vassar, and Henry Wells, among many others, donated significant portions of their fortunes to fund new colleges and universities, and they noticed that such ventures would quickly build endowments to rival those of the colonial colleges.[10] Northeastern students at institutions new and old picked up on the celebratory atmosphere of the era, and they created performance traditions—serenades and plays—that extended the tradition of burning textbooks to the stage. Communicating the sheen of expense, these student performances created a space where both men and women could act out their intelligence and good fortune.

Young women proved especially central to these new rituals. Metaphorically but not actually present in Bowdoin's Burials of Anna Lytics, their physical presence became the impetus for the math funerals of the Gilded Age. Male students from Columbia and Cornell increasingly decided to celebrate the end of certain requirements by serenading the female students of neighboring institutions. At these events, singing served as the communication of amorous intention, and banquets and dances threatened to replace textbook bonfires altogether. What they called the Burial of the Ancient or the Algebra Cremation did involve modified features of earlier math funerals, such as mock prayers and hymns, yet these features hid the male students' broader interest. The textbooks, so central to antebellum math funerals, slowly disappeared from the ceremonies altogether as young women took center stage.

The substitution of a woman for a book had much to do with the expansion of women's college education. Higher education for women, though controversial throughout the nineteenth century, had found many supporters in the antebellum South and Midwest. North Carolina's Salem College emerged shortly before American Independence, and over the following seventy-five years, it was joined by Antioch College (in Ohio) and Hillsdale College (in Michigan). Such "female colleges," as they were then known, had much in common with other institutions for women, including schools, seminaries, and institutes. In fact, as historians Thomas Woody, Kim Tolley, and Andrea

Radke-Moss have shown, these different names for institutions hid nearly identical frameworks: similar curricula, endowments, activities, and buildings.[11] The name "college" meant little for the reputation of a women's institution, though the striking popularity of Vassar College in the 1860s has perhaps confused the historical record.[12] In emphasizing (perhaps overemphasizing) its break with previous trends of women's higher education, Vassar College did allow for novel developments in college culture, reflected in the changing nature of math communication.

The roles of women in these celebrations reflected their assumed places in higher education generally. The female students at Rutgers Female College (in New York City) and Wells College (in the Finger Lakes region of New York State) acted only as audiences when young men traveled to sing their academic songs at them in the 1870s and 1880s. Like the earlier processions and parades, these so-called serenades relied on an assumed, unidirectional transmission: jokes about college knowledge told to the assembled, silent masses. The men who came from Columbia and Cornell did not treat these women as participants in their (usually prohibited) celebrations and denied them access to that particular aspect of college life at their institutions. Not maliciously exclusionary, they also did not include the students from Rutgers Female and Wells, except as passive audiences. In doing so, they resisted the ideal of such institutions: the ideal that women at Rutgers Female and Wells were college students, too, who had the same interests and abilities as male students at neighboring universities.

On a collegiate stage, the women who attended Vassar did take on strikingly new roles, ones that encouraged young men at nearby universities to accept them as college students too. Starting as early as 1871, they tried their own version of the textbook burial, which they called the "Trial of Trigonometry." Taking its inspiration from the active celebrations of Bowdoin and other colleges, it made the textbook burial legal and literary. Expanding on the document, Vassar students of the 1880s and 1890s wrote plays in which an actor representing mathematics met an unfortunate end. Many, such as the widely reviewed *Mathematikado* of 1886 (see figure 5), took their inspiration from existing works, into which they inserted mathematical jokes and characters. Their performances of "Trig Ceremonies" won wide acclaim not only for their originality but also for their college humor. Making jokes similar to those made by male peers, Vassar students showed that they too could understand and make fun of college subjects. At a time when some professors at men's colleges scrutinized women's performance to try to prove the infeasibility of educating women, these productions served as an important argument for students. By keeping a long-standing college tradition, albeit with some modifications, Vassar students appeared the right sorts of women to join the communities of college men.

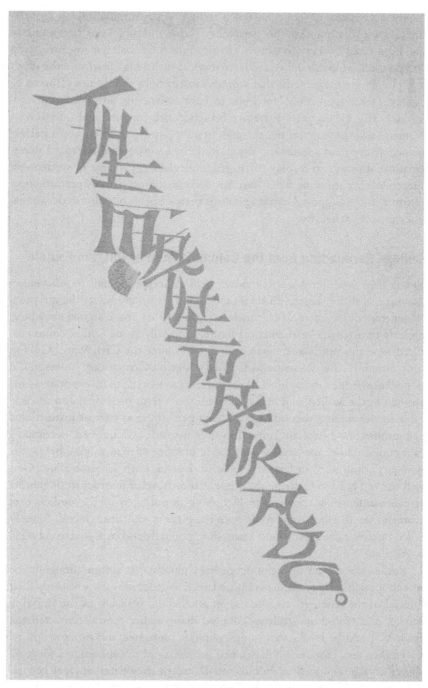

FIG. 5  Cover of *The Mathematikado* (1886). Personal collection.

The acceptance of "college women" relied on perceived academic abilities broadly and math abilities in particular. While the subject of trigonometry had been taught at certain women's institutions for decades, many Americans still doubted women's abilities to learn math at such a high level, and the 1870s saw popularized arguments that women's institutions should not claim to be colleges, since they could not hope to have sufficiently advanced courses.[13] Vassar's Trig Ceremonies therefore breached the controversial question of women's educational abilities through literary parody and humorous performance. More than spectators, Vassar students incorporated advanced mathematical allusions to display privileged, collegiate knowledge in student-led spectacles. The sheen of the Gilded Age decorated these parodic performances not just in the supposed extravagance of their displays but also in occasional assertions of exclusivity.

## College Serenading from the Columbia Men at Rutgers Female

Young men at Columbia and Cornell held elaborate funerals in postbellum America, and the changes in their ceremonies circa 1880 invite the question, What roles did women play for male "mourners"? As I have argued elsewhere, appeals to a female mathematical figure, especially Anna Lytics, communicated new conceptions of death and memory after the Civil War.[14] Columbia's Burials of the Ancient and Cornell's Algebra Cremations, however, did not acknowledge a metaphorical role but instead worked to incorporate actual women in the audience of *college serenading*—a term used then and since to refer to the singing that college students performed as part of institutional ceremonies. As historian James Lloyd Winstead has analyzed, serenading also connoted the medieval performance practice of men singing below the women's windows.[15] In fact, the male students at both Columbia and Cornell worked hard to re-create the scene as they traveled in order to sing under female students' windows. The increasing importance of the audience of women therefore had more to do with the place of academic (mock) funerals within performance traditions: serenading generally and college serenading in particular.

Both senses of serenading undergirded Columbia's tradition, though it was not about math: there, students burned ancient geography texts instead. Their "Burial of the Ancient" emerged as an antebellum offshoot of the Burial of Euclid, and Columbia students adopted many of the central characteristics from Yale and beyond: puns, songs, parades, costumes, and other props. At Columbia, too, "the textbook deemed most hateful to sophomores was consigned to flames amidst elaborate ritual," except they hated ancient geography instead of Euclid's geometry.[16] In addition to classes in Greek and Latin, Columbia required its students to take a class in which they memorized the changing cities, towns, and territories of ancient Greece and Rome. Though such classes were not unusual in the early nineteenth century, as educational historian Kim Tolley has reviewed, Columbia's ancient geography continued to appear on the list of requirements well into the 1880s.[17] Its widespread hatred emerged from the sense that such a subject could not be relevant to their current or future lives, not even in conferring mental discipline. The emerging tradition lasted at least thirty years, variously named the Burial of Ancient, the Burial of the Antiquities, or the Perideipnon (funeral banquet).[18]

The inaccessibility of ancient geography at Columbia also stemmed from the perceived remoteness of its central books: translated handbooks of Greek and Roman antiquities from the Danish philologist Ernst Friedrich Christian Bojesen. A scholar of Aristotle, Bojesen's interests in Greek music brought him to update the existing introductory textbooks about Greek and Roman society for his grammar school and academy students. His textbooks encouraged a rich imagining of ancient life with sections not only about politics, colonies, states, tribes, and wars but also about law, religion, family life, scientific thought, diet, fashions, and drama. It also set new trends: it was organized by country and included vast appendices about the customs common to all regions throughout a given empire.[19]

These qualities made the textbook more approachable for young students, but those at Columbia College found it literally foreign. Their *Manual of Grecian and Roman Antiquities* acknowledged not only the Danish roots of the new educational system but also the American need for intermediary German translations to construct an appropriate edition in English. Even though the manual also noted the importance of Greek and Latin texts for its production, students felt linguistically distant from the material: after all, they were learning about ancient life in an English translation from German, which came from Danish, which consulted Greek and Latin.[20] Moreover, the textbook encouraged students to turn the reaction into something dramatic, something performance-based, through its sections on Greek music education.[21]

In their burials, then, Columbia students attempted to assert the importance of local geography instead of a Danish interpretation of Greek life. Starting in the 1860s, assembled sophomores processed up Fifth Avenue to the college green, where they would gather to share mock prayers, poetry, and speeches about their hatred of Bojesen and his book. After throwing *Grecian and Roman Antiquities* on the bonfire, perhaps attempting to capture the ashes in an urn purchased for the occasion, they would drink large quantities of beer.[22] The burial committee, as usual from Yale and elsewhere, communicated the windings of the parade route, the proper approach for the bonfire location, and (importantly) the designated place for drinking, and these directions took on new meaning in a ceremony ostensibly about geography. The construction of the Burial of the Ancient asserted the importance of New York locations for the students, and it thus incorporated implicit (as well as explicit) arguments about the irrelevance of ancient civilization for their modern, urban lives.

New York City's geography proved especially central to Columbia's Burial of the Ancient because the college had recently moved. Founded as the fifth colonial college in 1754, King's College (as Columbia was then called) lost much of its property during the Revolutionary War, when its library was looted and its sole building was requisitioned for use as a military hospital for first American and then British troops. The resulting Columbia College (renamed for the new nation) remained in its aging building until 1857. By that time, the board of trustees elected to use the college's slowly increasing funds to construct a new campus, and ultimately the students and faculty moved into new Gothic buildings on Forty-Ninth Street between Madison Avenue and Fifth Avenue, as documented by Columbia historians.[23] Following the trends of some other colonial institutions, Columbia's architects did not provide lodgings for students; they needed to travel from nearby boarding houses or family homes farther away. The movement from Park Place to Forty-Ninth Street therefore changed their journeys to class significantly.[24] The Burials of the Ancient incorporated a hatred of geography at a time when students had to generate new ways of navigating the city.

The incorporation of the women of Rutgers Female College emerged through the institutions' relocation and resulting proximity. Shortly after Columbia moved between Madison and Fifth, the faculty of Rutgers Female Institute decided they too should move.[25] Fearing that prospective students might be lured up the Hudson to the new Vassar College, they remade their institution with a new curriculum, new building, and new name. Theirs had been a recognized New York City institution for forty years, but the faculty worried that the title "college" increasingly mattered for young women contemplating higher education. In order to receive the new name from the New York State Regents, they needed to show improvements, which they did through both academic and structural modifications. Following the example of Vassar College's trustees, they commissioned a grandiose multipurpose structure to house faculty, students, classes, religious services, reference collections, and staff.[26] The chosen site for the new building was Fifth Avenue, and the resulting proximity inspired college serenading.[27]

The Rutgers Female relocation coincided with the resurgence in college singing following the Civil War. At colonial colleges (perhaps Harvard), students had begun singing unofficially at the same time as official college events, usually commencement. College serenading had become a recognized collegiate practice by the 1720s–1730s, according to historian James Lloyd Winstead, and it had spread to Columbia soon after.[28] There, unofficial, though formalized, singing continued up to the 1860s, when it merged with the Burials of the Ancient. Sophomores did burn Bojesen's textbook during the years of the Civil War, except their official histories claim the celebration in the winter of 1864 as the true origin of the singing tradition. What they called their "simple, yet pathetic" ritual became more ornate after the war officially ended: incorporating urns and (in two cases) strolling German bands.[29] Increasingly, music, especially singing, reflected the wealth of their celebrations, and they found an audience in the students of Rutgers Female College.

In a postwar context, the assertions of American history in the Burials of the Ancient implicitly criticized Bojesen's Greek (and Danish) contexts. In 1879, for instance, the class of 1881 chose to parody postwar memorialization. On Decoration (Memorial) Day, "a day on which millions, all over this vast Republic, had been decking with garlands the graves of slaughtered heroes," they decided to "go through once more, sad, crematory exercises over the grave of slaughtered Bojesen."[30] According to the account in the student newspaper, the postwar students had been meeting annually at the war memorial to General William Jenkins Worth, not of the Civil War but of the War of 1812 and the Mexican-American War. There, they began the procession in "platoon[s]," with pallbearers, a band, and torchbearers following behind. Their pseudo-militaristic march began to sing shortly after, and they stopped to serenade the students of Rutgers Female College "as usual." Ending at the college grounds, they burned and buried their book, amid poetry and speeches, before they "fell into line" for a final march for beer. Throughout, their military language and iconography, including their plan for the burial to coincide with Decoration Day, placed the ritual in a distinctly American context. As they objected to Bojesen's extreme distance from their lives, the Burial of the Ancient emphasized the importance of American military events and sights through their jokes about them. The ritual implicitly communicated students' nostalgia for antebellum life.

The later tradition of placing the Burial of the Ancient within Anglo-Saxon history also harkened back to an imagined antebellum time. As Greek history and geography waned in popularity, Columbia's faculty began to incorporate Anglo-Saxon references and readings into their classes instead. For a time, in the 1870s, both subjects appeared, which resulted in classes burning and burying Bojesen's ancient Greek manual amid Anglo-Saxon and then an Anglo-Saxon reader amid ancient Greek references. Nicholas Murray Butler, long before becoming Columbia's president and winning the Nobel Peace Prize, helped to redesign the event and "to consign the Anglo-Saxon Reader to the flames."[31] Such references communicated the exclusivity of collegiate knowledge and, in particular, the moment when students were expected to know both ancient Greek and (mock) Old English. Moreover, the ceremonies featured their sense of the romance and excitement of the Middle Ages. In bringing the historical references forward in time, the students made them incrementally more relevant, except their funerals still did not include many features of modern, urban life. A historical setting, albeit an Anglo-Saxon/Greek one, remained essential, as the students romanticized life in the past.

Columbia students' imaginings of "Rutgers girls" were likewise nostalgic: in combining medieval and antebellum contexts, they especially communicated a desire for the return of earlier gender roles. According to an alumnus of the City College of New York (CCNY), Rutgers Female students were not incorporated into Columbia's ceremony except insofar as they were silent, romantic objects. Lewis Sayre Burchard, when a student at CCNY, joined the Burial of the Ancient with some friends, and he noted the Rutgers Female students' centrality. But the "Rutgers girls" were merely "fondly-imagined," not actually present, as the "Columbia boys" only saw the outside of their building: "the picturesque round-towered Rutgers 'Female' College buildings opposite the Reservoir."[32] The construction of the whole ritual as serenading, Burchard suggested, relied on the women to be distant from the male singers, merely "fondly-imagined." Such names and actions placed the Burial of the Ancient within a tradition of medieval courtship, which partially facilitated the rest of the medieval references surrounding the encounter.[33] Columbia students imagined not only their female audience but also the historical setting in which they believed the practice originated. Such envisioning facilitated their nostalgia, especially for an imagined time when sex segregation relegated women to the place of unseen, inaccessible romantic objects. Women, even from Rutgers Female, filled the daily urban environment but not these midnight escapades of the "Columbia boys." The Burials of the Ancient therefore communicated how, even when Rutgers Female students were not present and only imagined, they fulfilled a historic role for women: silently appreciating male performance.

## Serenading Math from the Cornell Men at Wells

Cornell's Algebra Cremation, unlike Columbia's traditions, involved implicit historicism and explicit incorporation of young women. Beginning in 1882, many young men from Cornell University engaged in what they came to call their "Excursion and Cremation."[34] Processing to the pier on Cayuga Lake, they boarded a large boat, disembarked at Wells College in Aurora, and engaged in the "usual" burial rituals there: singing, toasting, drinking, and occasionally burning the book of their latest math class. Though some ceremonies happened at village restaurants, usually the Aurora Inn, the attraction of the location came from seeing the young women enrolled at Wells. Despite the presence of female students at Cornell, which was coeducational at the time, male underclassmen still chose to "serenade" the "girls of Wells" nearly annually from 1882 to 1889, and the Wells students, unlike the ones at Rutgers Female, did appear at their windows from the first year.[35] The construction of the whole student ritual as historical, perhaps even pseudomedieval, emerged from the language used to describe the affair: the "excursion" was an excuse not only for "serenading" but also for "banquets" and dancing.[36] Such features connoted celebration and festivity amid a sense of historical, mathematical performance.

Appeals to history, even such slight ones, stood in stark contrast to the rest of Cornell University, constructed to seem innovative and modern. Conceived during the years of the Civil War, its proposal came as a response to the Morrill Land Grant Act of 1862, promising nearly a million acres for the founding of Union universities to focus on agricultural sciences, industrial studies, and military tactics. Ezra Cornell of Ithaca and Andrew Dickson White of Syracuse, despite initial differences, proposed the creation of a university with the federal grant and an initial endowment from Cornell's fortune in the Western Union Telegraph Company. After some waffling from the New York State Senate, the construction of college buildings began, and by 1868, 412 young men enrolled—the largest incoming class an American institution had ever seen.[37] Ezra Cornell in these initial years set the standard for the university, saying that it should be the place "where any person can find instruction in any study," except the Morrill funding led to its privileging of scientific, medical, and technical pursuits.[38] It had connections to telegraphy from its founding, a "water cure" facility that served as the initial dormitory, and in the year when students began their Algebra Cremation, buildings that started to be electrified through the water-powered dynamo. In terms of academics, too, Cornell of that time featured a range of offerings in the sciences, agriculture, engineering, and eventually medical studies—all linked through mathematics.[39]

In the varied curricular environment, math remained central, particularly through the efforts of the Cornell faculty in that department. Students in all courses had to take at least algebra and trigonometry, though students in architecture and all fields of engineering (civil, electrical, and mechanical) needed additional, advanced instruction. Early advocates of the elective system, Cornell and White had encouraged the differentiation of college courses, but some studies, they felt, proved necessary for all. Along with military drill, gymnastic exercises, and hygiene, algebra provided a common experience for students no matter their declared pursuits.[40] The faculty serving math maintained that their algebra represented "advanced algebra" as well, because it was also present on entrance requirements. Having difficulty finding a recent (modern) textbook that framed algebra in a university context, they wrote their own. Rife with examples and applications, their *Treatise on Algebra* reflected the place of math at Cornell. Administrators and students alike began to refer to it not by its title or authors but by an abbreviation: "Profs. Oliver, Wait and Jones" on the textbook's title page became "O.W.J."[41] In increasingly theatrical ways, these math professors became representatives for their department and stand-ins for the university's requirements.

O.W.J.'s *Treatise on Algebra* also introduced most Cornell underclassmen to historical study. As Yale president Jeremiah Day had earlier realized, history provided an opportunity to showcase applications of logic, arithmetic, and simple algebra.[42] Similarly, Oliver, Wait, and Jones illustrated both logical equivalence and subtraction through the example: assuming that "the battle of Salamis was fought 480 B.C. and that of Waterloo 1815 A.D.," then "Salamis was fought 2295 yrs. before Waterloo" and "Waterloo was fought 2295 yrs. after Salamis."[43] Though examples involving money and temperature appeared more often throughout the book, considerations of chronology and historical dates peppered many sections. Cornell had a variety of classes about the historical life and times of various past empires, except the enrollment in such classes did not approach algebra's, because of the founders' preferences. In short, algebra classes provided the place where most Cornell students would consider past battles of ancient and modern history.

At the time, Cornell's Algebra Cremation combined historical references with distance from the university. For the students, algebra and history appeared related, as the examples from their textbook indicated. Algebra's "death" and memorialization encouraged the conflation of these two even further, as underclassmen gathered to imagine a "past" for a personified mathematical foe. The remote location of Aurora, so unusual with respect to other math funerals, provided a fitting setting as a small village near their electrified university, and these comparisons inspired considerations of math's earlier time. Aurora, too, provided a place clearly removed from the university where students could proclaim their hatred for the mathematics department and the broader institutional requirements.[44] Their "O.W.J.," after all, represented Cornell as a whole, and their speeches expressed an implicit reluctance to criticize the university on its own grounds.

Wells, in fact, had rejected Cornell. In 1866, philanthropist Henry Wells expressed his wish to support women's education through his fortune made through the successes of the Wells Fargo and American Express Company. At first, Ezra Cornell approached him about the possibility of funding his university's anticipated experiment with coeducation, but Wells declined, saying that he wanted to found an institution for women that would "promote a higher standard of moral and intellectual culture" than they could find at "ordinary village and town institutions."[45] The public rejection of Cornell, particularly with the possible conflation of his enormous university with small "town institutions," entered regional lore after the opening of Wells Seminary in 1868. There, Wells and his faculty claimed, women would achieve "moral and intellectual culture" befitting their future roles as wives and mothers, and they emphasized the difference between their goal and Cornell's to create a place "where any person can find instruction in any study." Their domestic atmosphere and moral-religious subjects made clear what was missing in a large, industrial university. When women started attending Cornell classes in 1870, and when the university constructed a separate space for women's lodging and classes in 1875, Cornell faculty did follow the model of Wells, and some even talked about Wells College as a "sister college" at that time. A grudging cooperation between the two institutions remained into the 1880s and beyond.[46] In bringing their Algebra Cremation to Wells, then, Cornell underclassmen did not fear reprimand and, in fact, expected to find people who would join them in their antics as they poked fun at their university.

The Wells students, however, only participated in the role of an appreciative audience, and the planning of Cornell students limited any further involvement. The upperclassmen periodical, the *Cornell Era*, parodied the relationship through creating a "Programme" from the second annual "Sophomore Excursion." After approximately fifty Cornell underclassmen disembarked at Aurora, they happily processed down the town's short Main Street to Wells College, where the "inmates" were "picturesquely grouped about every available point of observation in anticipation of their arrival."[47] The program then proceeded as follows for part I:

Cornell yell [from] Cornell students
Applause [from] Young ladies of Wells College
Selection . . . Band
Cornell yell . . . Cornell students
Song, "Spanish student" Cornell students
Enthusiasm . . . Young ladies
Cornell yell . . . Cornell students
Lingering expectancy . . . Everybody
Cornell yell . . . Cornell students
More enthusiasm . . . Y.L. of W.C.
Song, "Far above Cayuga's waters," Cornell students
Rapturous applause . . . Young ladies.[48]

Part II went in a similar manner:

Cornell yell [from] Cornell students
Bouquets ad libitum [from] Young ladies
Three cheers for a sister college, Cornell students
Overwhelming applause . . . Young ladies
Cornell yell . . . Cornell students
Presentation of programme of the evening to Young ladies . . . Small Boy delegate
Cornell yell . . . Cornell students
Pause
Cornell yell . . . Cornell students.[49]

By way of conclusion, the anonymous authors of the article joked that such a program showed the necessity of "vocal music" classes at Cornell, since they clearly showed "musical capabilities" when "abroad on a serenade."[50] It also demonstrated the limited participation of the "Young ladies of Wells College" who could only show "applause," "enthusiasm," "more enthusiasm," "rapturous applause," and "overwhelming applause." Not able to leave their college, they remained "inmates" who only acted in ways to indicate their appreciation of their male guests.

According to such a program, which only showed an idealized version of actual events after all, Cornell's Algebra Cremations also served to reassert historical gender roles through traditions of theater. As in medieval traditions of serenading, reified in college settings at Columbia and elsewhere, women acted as merely an appreciative audience for male performance. As historian James Lloyd Winstead has observed, even the physical presence of female students for late-century serenading caused a news sensation.[51] In the context of late-century college education, such events therefore emphasized the importance of gender through the reification of male traveler-guests and female resident-hosts. As in the highly idealized middle-class family of antebellum America, the men returned to women's domestic realm after work. In women's explicitly domestic spaces, men performed the results of their professional development, expecting appreciative, though not substantive, reactions.[52] The incorporation of algebra into the Wells course of study did not appear in the Cornell students' plans, nor would it in their framework. As female resident-hosts, as idealized wives and mothers, the "Y.L. of W.C." could only welcome and applaud, not show their understanding of math references or their abilities to poke fun at the subject either.

The students engaged in Columbia's Burials of the Ancient and Cornell's Algebra Cremations did not see the irony in imagining a time before the widespread expansion of women's collegiate education. Both of their ceremonies incorporated nearby women's colleges as central to their revelries. Female students, whether real or imagined, were featured as romantic objects and appreciative audiences for their performances of "hatred." Columbia and Cornell traditions therefore hinged on the presence of Rutgers Female College and Wells College. While the men's actions communicated nostalgia for historical gender roles, their ceremonies necessarily had to follow the reforms that expanded women's higher education. The student traditions at Columbia and Cornell reflected a broader tension between imagining an academic past and envisioning its future. Theatrical conventions facilitated the nostalgia, and they also provided a way for female college students to speak back.

## Women Embodying Math at Vassar

The concurrent developments at Vassar College presented an alternative role for educated women, not just as an audience but also as instigators. Students for generations had been performing their "hatred" of mathematics, as they named it, but Vassar students took the tradition to the page and the stage. The drama, in these cases, resulted from the framing of the tradition as a battle between two opposing foes: a "college student" and a personified figure of mathematics. The stories, first written and then acted, dealt with the confrontations of these two characters, presenting the ways that mathematics made life difficult for the college student and the ways that the college student ultimately vanquished mathematics. Often, though not always, the figure of mathematics was led to throw itself on a funeral pyre as a penultimate scene of the composition, leading to the ultimate rejoicing of the rest of the characters. The Vassar version of the textbook burials began as a literary exercise in 1871 but quickly became a dramatic performance, recognized not only at the college but also at a host of other academic institutions by the mid-1880s. These "Trig Ceremonies," as Vassar students called them, extended the tradition of embodying math from Yale and beyond.[53] It also communicated ways of being college students, which proved an especially complex task in an academic environment where "female college student" often represented a contradiction in terms. Through parody and humor, Vassar's Trig Ceremonies therefore exemplified the prospects not only for women's math abilities specifically but also for women's ways of being college students broadly.

Vassar's first Trig Ceremony in 1871 established a clear link to previous rituals enacted by male college students. A few members of Vassar's fifth class of sophomores created a pamphlet called the *Trial of Trigonometry*. Produced on a local steam printer, the pamphlet was laid out not as a script but as a narrative with dialogue.[54] Written by three named members of the class of 1873 and in the voices of two others, it featured the arguments for and against burning the book. A. Skeel, for the prosecution, began with a parody of Shakespeare's *Julius Caesar*: "We come to burn our Trig, not to praise it. The evil which books do, lives after them; the good is rightly buried with their bones. So let it be with Trig."[55] E. Weed, for the defense, reminded the sophomoric jurors of their first encounters with trigonometry and then recited a monologue in the style of *Hamlet*: "To burn, or not to burn, that is the question."[56] Despite Weed's appeals to "love" and memory, the Vassar jury did decide to burn their book, which led to the antics typical of male rituals. In the pamphlet's story, the members of the class of 1873 processed to the Vassar Farm, listened to a classmate's oration about the need to burn more books, and sang a popular song with modified lyrics that joked about their "escape" from trigonometry.[57] In their adoption of a recognizably male tradition, the sophomores implicitly claimed that Vassar had equal college status to the other (men's) institutions that burned their mathematics books. However, Vassar students could only completely enact the ritual in fiction; as women, they faced increased scrutiny.

Vassar's collegiate status had proved problematic since its founding. Matthew Vassar, who had made his fortune brewing beer, decided in the late 1850s that he wanted his hometown "monument" to be an institute for "the enlarged education of women."[58] Since Vassar had family connections at the all-male University of Rochester and the Cottage Hill Seminary for women, he began by consulting teachers at those institutions. In their conversations, Vassar and his friends became convinced that the proper way of communicating "enlarged education" was to charter the institution as a college. At the time, only two women's institutions claimed similar titles in New York State: Elmira College and Ingham University, both in the western part of the state. (Rutgers Female College had not yet requested a new charter, and Wells College's founding occurred later.) Vassar's proposed college, located north of New York City near the Hudson River, satisfied a perceived regional need.[59] Despite the presence of clear precedents inside and outside the state, the college's officials quickly adopted the language that such an institution was an "experiment," the first institution of higher education of its kind for women.[60] The excitement of Vassar's rhetorical stance not only attracted applicants, but it also invited Americans to imagine the project as liable to fail. Even before students arrived, Vassar's officials posited the "Female College" as precarious.

Because of such language, early Vassar witnessed tense discussions about the role of gender in higher education, and the problem of practically implementing such a conflicted system caused many early faculty members to leave. Some had experience teaching at other women's institutions: Charles Farrar (mathematics, natural philosophy, and chemistry) had taught at Elmira College; Maria Mitchell (astronomy) at Cyrus Pierce's school; and Hannah Lyman (lady principal) at New England schools.[61] Of the seven remaining, most left promptly because of their discomfort with Vassar's "experimental" atmosphere. Though Alida C. Avery (physiology and hygiene), Henry Van Ingen (drawing and painting), and college president John Raymond (mental and moral philosophy) remained after the college's first years, Henry B. Beckham (rhetoric, belles-lettres, and the English language) stayed for only one; William I. Knapp (ancient and modern languages) and Edward Wiebé (vocal and instrumental music) for two; and Sanborn Tenney (natural history) for almost three.[62] Tenney, like his colleagues, at first embraced the project of designing curricula equivalent to men's and yet "suited to woman's physical and intellectual wants."[63] But the tensions inherent in such a pedagogical framework ultimately led to Tenney's sudden departure for all-male Williams College, as his colleagues had previously left for normal schools, Yale, and the University of Chicago. Balancing claims of equality and difference had previously led to the renaming of the institution too. The New York State charter read "Vassar Female College," and that sign appeared at the top of the main college building. But the trustees decided to drop "Female" from the name before students arrived, leaving a noticeable blank that remains today.[64] Some early professors, such as Tenney, worried about the wisdom of the institution's pedagogical choices, and faculty turnover did not provide curricular stability.

Early students argued that the administrative rhetoric of Vassar's "experiment" trivialized their educations, and some explicitly lobbied to simply be called college students instead. Mary Harriott Norris, a member of the class of 1870, remembered how her education consisted not just of classes but also of feelings of pedagogical instability. She generalized that students always felt like they were being watched, as "the vicarious expression of what was considered a most hazardous experiment."[65] Alumna Frances Wood, too, remembered how "it was impressed upon the whole family that the higher education of women was an experiment, and that the world was looking on, watching its success or defeat."[66] In such moments, the college was expected to stand together as a "family" before the "world," despite student misgivings about their role in the enterprise. Not treated as the experimenters but only as scientific objects, they had little control.

Ellen Swallow Richards, the later founder of home economics, argued in her correspondence that Vassar students lacked the vocabulary to talk about their roles. Labels assigned to them proved insufficient. Instead of "school girls" at a "school," Swallow suggested they be called "students" or "hard workers" instead, as they had to do well in their classes just as men did. Her reactions succinctly noted the connections between their identities and Vassar's, as she wrote, "Vassar *College* is occupied by *students*, not school girls."[67] In other words, the institution's identity as a "college" was upheld by a broader perception of its pupils as college "students," regardless of whether they were "girls" or not.

At the time, the 1871 *Trial* emphasized that the sophomores were college students, too, not "school girls." The extension of the practice of textbook funerals made an implicit argument about Vassar's status, forging a tie to traditions at men's institutions. In the pamphlet's appeals to the student body, their collective identity was recognizably collegiate: "a class" of sophomores, a "Sophomore Class," and "we" who imagine "our Senior year."[68] Moreover, in their story, both prosecutor and defendant generated phrases that ignored the gender of the students involved, as in their references to each other by a first initial and last name (i.e., A. Weed). The overall effect communicated the identity of a group of college students, ones who did not modify that label through considerations of gender. Such an attitude reflected the trustees' early deliberations. The 1871 Trig Ceremony, though unofficial, even rebellious, did still reinforce the official message that Vassar College was on par with male peers.

The *Trial's* character of mathematics, too, had names that related to the broader question of women's higher education. Though given the nickname "Trig," it appeared as one of many "books" and thus received the pronoun "it."[69] Only partially personified, it acted on and through the students, and the drama of the pamphlet came from the anticipated actions of the student protagonists. Much later, Trig's gender in the "Trial" was revealed in the last line of the last page. Once burned, it could not harm the students, as they sang, "We've escaped the wretch's thrall / Never more *he'll* bore us."[70] The last page of the pamphlet therefore communicated how, in an academic landscape, appearing genderless equated with appearing male. Within the context of the disappearing "Female" sign, the twist suggested possible implications for the college and student body as a whole.

Shortly after the *Trial*, women's higher education generally and Vassar's collegiate status in particular became a topic of national and international debate. In 1873, the physician and former Harvard professor Edward H. Clarke published arguments against the ideal of equal education in the now infamous volume *Sex in Education.*[71] As many historians have studied, Clarke used medical case studies to support his claim that women should not be educated according to men's standards. Such attempts at women's higher education, according to Clarke, could lead to problems in menstruation, resulting in sterility, masculine behavior, insanity, and death.[72] Clarke particularly worried about coeducational institutions, which he kept anonymous, but he unexpectedly named one college where he felt the faculty and parents had enforced policies harmful to student health: Vassar College. Though not in fact coeducational at the time, Vassar represented the problems of the broader academic landscape for Clarke: pushing a "complicated" educational system on girls who also had to develop what he called a "complicated reproductive mechanism."[73] To Clarke's surprise, his book sold very well in the United States and abroad, and it inspired anxieties in college towns throughout America, especially in college women themselves. Through it, Clarke spread his belief that women should not pursue a collegiate education, especially at Vassar College. Through their Trig Ceremonies, particularly through their theatrical versions of math, Vassar students argued back.

At first, the Trig Ceremonies, as dramatic performances, eschewed problems of the masculinity of the "college student" label through the bodies of female performers. Some, such as the *Trigonometrical Bluebeard* of 1881, contained an unabashedly feminine character to represent the entire class.[74] In their version of the play, Trigono *Metry* appeared as the threatening Bluebeard, and Sophie M. Ore, according to a student news report, "a young lady desirous of some one to appreciate her."[75] Despite her familiarity with the Bluebeard story, Sophie still decides to wed Trig, receiving a triangle earring as a symbol of their union. An unexpected hero of "Sam Ester" saves Sophie from

Trig's murderous knives, and the play ends happily as Sophie M. Ore "drops" the triangle, "the tinkling symbol of her woe."[76] In their version, the personified mathematics remained a threatening male, as in the *Trial* ten years earlier, but the drama of the production relied on the ways that the representative of the class acted young and feminine: naive, impressionable, foolish, and incapable of finding "appreciation" except through others. In the central marriage and adornment, too, Sophie's actions spoke to a subordinate role, as she took the symbol of her husband and incorporated it into her dress and identity. Such actions portrayed the sophomore class, in personification and then in sum, both as clearly college students ("Sophie M. Ore" after all) and also as clearly feminine. The play did not present the identity of "female college student" positively, as Sophie's scholarly knowledge did not prevent her from succumbing to a doomed marriage. Not agreeing with Clark and his ilk, in the *Bluebeard*, women's scholarly endeavors lead to a poor choice but then an ultimately happy ending.

After *Bluebeard*, Vassar students usually cast themselves as men to counter Clarke's claims. The Trig Ceremonies served as a place where students could show off, proving to members of the college "family" and female guests that they could master college-level mathematics, not to mention Latin, botany, literature, logic, and economics. However, unlike public examinations (described in depth in chapter 5), these events became a place where Vassar students explicitly acted like men. After 1881, the Trig Ceremonies featured a central drama between a *female* mathematical figure and a *male* college student, named only for the sophomores' anticipated graduation year. As literary critic Laura Kasson Fiss and I have investigated elsewhere, the 1886 *Mathematikado* starred a young man called Ayty Ayt (88), torn between his new love for Trig Trig, his old love for Latisha (Latin), and his cordial dislike of Bot Ah Nee (botany).[77] In addition, in an 1889 retelling of Dido and Aeneas, a male "91" appeared opposite a "Queen Trig," who threw herself on a funeral pyre at the end of the production. In an 1890 retelling of Orpheus and Eurydice, a male "Ninetytus" (92) attempted to escape trigonometrical demons with his betrothed. The Trig Ceremony of 1891 imagined a version of *Pilgrim's Progress* in which a male "Ninety-three" had to save a young woman from the menacing mathematical giant and also had to escape the flirtations of characters named "Literature," "Logic," and "Economics." The 1892 ceremony featured a "breach of promise" case between "Miss Trigonometrie" and "Mr. Ninety-Four."[78] Every instance featured a *male* representative of the class, and in each case, his actions stood in for the feelings of the whole student body: loving, hating, pursuing, and escaping mathematics. They could mock Clarke's worries about educated women, particularly through the avatar of a male college student.

The Trig Ceremonies' use of male protagonists mirrored Vassar students' collective experience of portraying men in their mathematics classes generally.

Their math textbooks consistently used male pronouns to refer to presumed readers. George Wentworth's *Elements of Plane and Solid Geometry* began by advising the student, "He should state and prove the propositions orally, using a pointer to indicate on the figure every line and angle named. He should be encouraged, in reviewing each Book, to do the original exercises."[79] *Chauvenet's Geometry* included such addresses within the text itself. After a proof of the theorem that "only five regular (convex) polyhedrons are possible," it stated, "The student may derive some aid in comprehending the preceding discussion of the regular polyhedrons by constructing models of them, which he can do in a very simple manner."[80] Though the way of referring to "pupils" or "students"—indeed, any collective noun—was common practice at the time, it echoed the ways that the women of Vassar naturalized masculine pronouns in college traditions, whether official or profane.

The humor of the Trig Ceremonies added additional complexity to these portrayals of gendered mathematics. The Trig Ceremonies allowed Vassar students to think through some of the ramifications of the dual pressures to represent themselves as both women (healthily female) and college students (conventionally male). But Vassar's Trig Ceremonies did so through the frame of jokes, providing a subtle way of calling attention to the paradoxical juxtaposition between masculine roles and the bodies of the female actors. Students' theatrical costumes of the time communicated the gender negotiation, according to Vassar chroniclers and performance scholars: not only at Vassar but also at many other women's institutions, actors wore male dress above the waist and traditional skirts below.[81] Though female students also played male characters elsewhere, as historian Barbara Miller Solomon has discovered, few joked about the experience until much later.[82] At 1880s Vassar, through the humor of their Trig Ceremonies, students found a way to ask the same question as Clarke: What is the role of the female body in (traditionally male) college education? By constructing a theatrical response to mathematics, Vassar students highlighted the complex considerations about where women fit into American college cultures.

## Theatrical Math at Vassar and Beyond

At the time, theater provided new possibilities—especially for women. Student productions provided a new occasion for showing off, especially showing off mathematical knowledge in oral communication. Though very few people from outside school and college communities could actually see women's performances, especially compared to their oral exams, many newspapers and even novels spread their stories, as we will see. Because of the reviews and broader media reception, Vassar's Trig Ceremonies as a whole—and the 1886 production of *The Mathematikado* in particular—subtly argued against public worries about women's higher education, showing female students could be healthy and funny—particularly about mathematics.

Male students did see themselves in Vassar's dramas when detailed reports of the 1886 *Mathematikado* appeared in student newspapers at men's colleges (Harvard and the Stevens Institute of Technology) and coeducational institutions where men dominated college journalism (Cornell). Unlike earlier Trig Ceremonies, which parodied Shakespeare, Milton, or Classical authors, *The Mathematikado*'s subject was a work from contemporary pop culture: *The Mikado*. Furthermore, *The Mathematikado* represented considerable literary effort: Vassar students not only replicated some of the plot of *The Mikado* but also rewrote the lyrics of eight songs in part or in full. When young men reported on the cleverness of these songs, they connected the mathematical modification of lyrics with their annual traditions of burning books, which featured similar components. Students at other institutions could see themselves in the male hero of the show and the broader project of finding creative ways to practice oral communication about mathematics.

Still, their articles—and reports from regional papers—communicated only partial acceptance of Vassar as a "college" and its residents as "students." The *Cornell Daily Sun* noted how Vassar's production would interest Cornell students because of the similarities between the Trig Ceremonies and their Cremations of Algebra, but it also called the performers "sweet girl graduates," not the usual "students."[83] The *New York Times* and the *Rome Daily Sentinel*, too, praised the originality of the students' parody while calling them "Vassar girls" throughout.[84] Only the reports from the *Stevens Indicator* and *Harvard Daily Crimson* referred to the performers as "students" without qualification.[85] When the Harvard paper engaged in scissors-and-paste journalism—borrowing its article directly from the *Poughkeepsie Eagle*—its description of the Trig Ceremony's performers as "college students" sent a strong message, as Harvard and its surrounding areas contained many of Clarke's supporters who worried that college was unsuitable for women.

The Trig Ceremonies also became part of the bigger project of legitimating women's colleges. As part of a feature on college life, the *St. Nicholas* children's magazine ran a summer issue with a lengthy explanation of "Festival Days at Girls' Colleges" from Wellesley College, Bryn Mawr College, Smith College, the Harvard Annex (later Radcliffe College), and Vassar.[86] Beginning with a stylized account of whispered conversations in an unnamed college's library, the article proceeded to give a firsthand account of Wellesley's Tree Day, a lengthy quote from M. Carey Thomas about Bryn Mawr's lantern ceremony, abbreviated explanations of Smith's Mountain Day and the Harvard Annex club system, and detailed histories of Vassar's Dome Parties (for astronomy) and Trig Ceremonies. Though the author, Grace W. Soper, was an early Cornell graduate, she paid particular attention to Vassar's traditions, calling them "unique and famous."[87] *The Mathematikado* proved the prime example for Soper, and she devoted a full column to its description, quoting most of the "Little List" parody. Throughout, Soper maintained that these traditions showed how women's colleges were places of "fun," not just "books," and how women had concocted pleasant alternatives to the "rude hazing customs of old days in young men's colleges."[88] Unusually marking "men's" colleges, Soper gradually called women's merely "colleges," as in her final address to the reader: "Other colleges have many bright evenings of recreation, about which you would like to hear, perhaps; but some day you will enjoy them for yourselves, after passing those dreaded entrance examinations."[89] Though the final remark did mention anxious exams, they were at the beginning, not the end, of studies. The effect of the whole was to address previous worries about women's education, making the featured institutions seem sites of gleeful college life.

Fictional accounts of student life at Vassar continued to use the Trig Ceremonies as a sign of collegiate status, even after the ceremonies stopped being performed. The winter of 1897 witnessed the last of the Trig Ceremonies, reflecting the faculty's "desire to simplify the social functions of the rapidly growing college."[90] In the final version—an 1897 parody of the Salem Witch Trials—the sophomore class as a whole passed judgment on "Goody Trig" and her assistants of Sin, Cos, Tan, Sec, Cosec, Cot, Vers, and Covers.[91] By 1897, the opportunities for dramatic performances had increased through celebrations of George Washington's birthday, tree ceremonies, Founder's Day, Halloween, and Valentine's Day.[92] There were also fewer public criticisms of women's education, and the college no longer needed to support a tradition that relied on student proclamations of hatred, even ones that sent a clearly collegiate message.

After their prohibition, the Trig Ceremonies continued to appear as representatively collegiate in fictional accounts of Vassar specifically and women's colleges broadly. Jean Webster's *When Patty Went to College* (1903), Charles Bolton's *The Harris-Ingram Experiment* (1905), and Julia Schwartz's *Elinor's College Career* (1906) aimed to make the college experience approachable and accessible for American girls, and they mentioned Trig Ceremonies in their characters' reminiscences, enthusiasms, and intrigues.[93] These specific mentions exemplified the whole books' effect of representing the fun of college life. Normalizing the college experience for women, these fictional accounts served to expand notions of the "female college student," as did popular texts and films of the later twentieth century.

While the Trig Ceremonies implicitly argued for the suitability of women's college education, Vassar's faculty made such arguments explicit. Professors Alida C. Avery and James Orton engaged in sustained campaigns against fears of women's overeducation.[94] Meanwhile, the actions of the mathematics faculty gave students hope for broader collegiate acceptance. Achsah Ely and astronomer Mary Whitney represented Vassar in the founding membership of the New York Mathematical Society, which became the American Mathematical Society, and Ely also represented the college at the second annual International Congress of Mathematicians, as has been noted by math historians Della Dumbaugh Fenster and Karen Parshall.[95] At a time when scientific societies regularly excluded women, Ely and Whitney represented not only Vassar but also college-educated, intellectually active women broadly, and their health and vitality exposed the errors in believing women could be overeducated. Vassar students, through their Trig Ceremonies, joined their faculty in performing women's roles in traditionally male institutions.

Men's colleges, too, followed Vassar in adopting theatrical outlets for their math funerals, and their students' productions occasionally retained the central tension between a male representative of the class and a female personification for mathematics. Amherst College's Burial of Mattie Mattix in 1883 included a parodic play about the students' suffering under a female Mattie, and Rutgers College's Burial of Anna Lytics in 1888 was an elaborate parody of *The Scarlet Letter* with a central trial that accused Anna of various puritanical misdemeanors.[96] In these cases, male students impersonated female mathematics, as in the case of Amherst student Clyde Fitch. As historians Margaret Nash, Danielle Mireles, and Amanda Scott-Williams recently uncovered, Fitch's female impersonation of Mattie Mattix presaged his later fame as a Broadway writer and director.[97] In short, theatrical math inspired actual careers, though not all of them mathematical.

At Amherst and Rutgers, having a male character as the representative of the class did not seem particularly surprising, given the performed gender of the students, as well as the literary and social conventions that assumed college men presented as male. Male students at ostensibly coeducational Cornell and Bates (in chapter 2) constructed math funerals to be a feature of exclusively male student life. In math plays, translating college conventions to the stage, students continued to perform the distinction of collegiate education in a public setting, and only Vassar's versions began to question the institutional frameworks that supported them.

For math students at many colleges and schools, in conclusion, actual theaters served as highly symbolic spaces in which actors, directors, writers, and other makers crafted new identities amid conflicting messages about gender, studenthood, and their relationship to learning mathematics. On stage, women could be men, men could be women, and anyone could take on the role of mathematics.[98] Generally, these events did not lead participants to devote their lives to the theater. Rather, it established bonds of studenthood through communicating success over math and over academics beyond.

The emerging math curricula partially reinforced the quotas that asserted men—especially white men—should be the dominant students, joining similar conventions from the earlier century. The blackboard, as in chapter 3, had assumed certain modes of oral communication from military schools, where only men attended. The earlier expectations for textbooks' reading, from chapter 1, similarly had framed math students and teachers as male; N. W. T. Root did not expect women to take off their shirts and join the mathematical gymnastics. As explored in chapter 2 and here, math funerals grew to practice these messages, incorporating women as passive audiences in many cases. But students at Vassar College, in taking to the stage, did find ways to assert that women could be college students by showing that they could learn college-level mathematics. In the changes to come in the twentieth century, such displays of math ability became less tied to literal stages, in part because of the privileging of written examination over oral examination. Still, written exams retained certain performative dimensions of math communication, particularly in their continued connections to stage fright.

# 5

# How Math Anxiety Became about Written Testing

• • • • • • • • • • • • • • • • • • • • • •

From the street and the theater, the students settled back into the math classroom at Vassar, Amherst, Rutgers, Cornell, and Wells, but just as their performances of mathematical success changed, so did the expectations for math communication. The show of learning math had new rules. In what math historians Peggy Kidwell, Amy Ackerberg-Hastings, and David Roberts have called the second period of blackboard use, communicative conventions were now deeply entrenched, creating a new lived reality for students from the Civil War Era to now.[1] To be successful in math meant calmly, accurately delivering an oral report as a military man would. Students' written communication at their slates was irrevocably replaced by oral communication at the blackboard. Such assumptions were so widespread that they were rarely questioned, except in some cases when women found the system stacked against them. As analyzed in chapter 4, the performative dimensions of mathematical speaking found new life in the theatrical traditions of serenading and student plays. Expectations for boardwork, as exposed in chapter 3, also suggested that students might experience a form of stage fright in front of their peers. The performative dimensions of math communication can be further explored in how math anxiety continued under new systems of written testing. Between the Civil War and 1900, American math once again returned to systems of written communication, in part as an attempt to standardize educational experiences throughout high schools and colleges and also throughout many sites around the country. The blackboard no longer acted alone in inspiring students' stage fright; so too did written testing.

Stage fright, in the next period of math communication, continued to be a historical analog of math anxiety. Though written communication did not involve going up to a stage in the same way that some forms of boardwork did, educators did act to expand notions of "student performance" to encompass written testing. Arguing for standardization, new educational researchers specified the importance of quantifiable performance measures, linking math, performance, and written testing with far-ranging implications. Leading to the late twentieth-century popularization of "math anxiety" per se, such trends in math communication encouraged stage fright, in part, by expanding what counted as a stage and what counted as performing mathematics.

Material tools of mathematics pedagogy again illuminate trends in knowledge communication—what it is possible to know or say in light of pedagogical technologies. These analyses build on Kidwell, Ackerberg-Hastings, and Roberts's grounding overview of what they call "standardized tests."[2] Though they have a less explicit periodization than for the blackboard, they note some important trends: the fractured relationship between school tests and college admissions tests from the 1830s to the 1880s; the movement toward the quantitative measurement of teaching from the 1880s to 1900; early tests of the variation in arithmetic teaching from 1900 to 1915; and the expansion of such testing (to individuals, to further math subjects, and to predicting student performance) after 1915.[3] Acknowledging these insights, the following remains for longer in the nineteenth century, investigating the early century reputation of public (oral) examination, the midcentury relationship between class testing (orally, at the blackboard) and professional (written) certification exams, and the late-century expansion of notions of "student performance" in advance of twentieth-century math anxiety. Adding to previous work about the material artifacts of math teaching and research, I moreover continue to argue for a shift in focus to student experiences.[4] Such a shift leads us to spend more time analyzing older trends in math communication.

In fact, the movement toward standardization—and the emphasizing of written testing that came with it—stemmed from the myriad changes American higher education experienced from the Civil War. As noted by historian Michael David Cohen, the wartime and postwar federal governments took a greater role in educational funding, management, and policy.[5] First, during the war and immediately after, the United States Congress passed acts funding certain types of colleges that emphasized certain fields: for instance, the Morrill-sponsored Cornell University.[6] Second, as the federal government expanded, it also established a new bureau, a new part of the federal government, to focus on education: collecting educational information, ordering it, and disseminating it. The Office (later Department) of Education encouraged the adoption and expansion of certain courses that seemed especially pressing for the time: military, industrial, and agricultural.[7] Lastly, federal initiatives joined state-based ones as demographic changes pushed high schools and colleges to accept more diverse students.[8] Written math tests, in part, developed as measures of students' differing backgrounds: by 1903, arithmetic scores could be compared by gender and ethnicity, in addition to home environment, school, and specific teacher. Federal involvement in educational funding, management, and especially policy had profound implications for the ways that math students experienced their classes.

Written testing also developed through industrial changes associated with the increased production and distribution of school supplies, from blackboards to pencils and paper. As curator Deborah Warner has analyzed, mid-nineteenth-century Americans began to patent "liquid slating" for the creation of blackboards out of practically anything: walls, boards, paper, cloth, globes, and so on. Eureka slating, excelsior slating, and many others appeared in educational magazines and were advertised in cities and towns through broadsides. Such innovations made possible the creation and growth of educational markets.[9] Slightly later, mass production allowed for the cheap distribution of steel-nibbed pens and artificially dyed inks. Graphite pencils began to sell cheaply enough for schools and colleges, and paper became less expensive through changes in chemical manufacturing, mass production, and experimentation with ingredients, from rags, cloth, and skin to grass and other vegetable matter.[10] Responding to blackboard markets, newer industrial changes pushed the corresponding development of written homework and written tests.

Math did have a special role in popularizing written testing, drawing the attention of educators throughout the country. From the 1860s to the years around 1900, teachers and newly minted educational researchers praised the quantifiable dimensions of standardized written testing, how it could reach large numbers of students with greater uniformity, greater accuracy, and greater effects than any previous model. As we will see, tests could expose inefficient teaching methods, measure students' learning, help teachers establish fair grading standards, or justify students' specialization/tracking throughout large numbers of schools and colleges. Though administrators and researchers rarely wrote about negative reactions from students and rarely reflected on the mathematical methods that justified their studies, they nevertheless debated the wisdom of written testing.

Learning math, in this way, has again intersected with expectations for performance and communication. Between the Civil War and 1900, none of the previous communicative trends went away. As in chapter 1, professors and teachers still asserted the importance of reading from math textbooks and, partially in doing so, still pushed the ideal of the disciplined, obedient math student. As in chapter 2, students still practiced their math lessons through increasingly prohibited math funerals and the like. The blackboard (from chapter 3) remained the classroom technology of choice, becoming increasingly overlooked as it became more ubiquitous—and yet still promoting a sense of stage fright in many students who had to follow rules for public speaking in class. Math plays featuring students (as in chapter 4) continued to be performed, even extending into the twenty-first century with the works of playwright Lauren Gunderson.[11]

Yet there were changes in math communication's continuity. New trends of standardization encouraged the questioning of previous orthodoxy, including mental discipline. Math funerals and their ilk became less practiced as more students enrolled with increasingly diverse interests. Most of all, written testing created new expectations for communication in math classrooms—making math less regularly about boardwork and making math less obviously performative. Still, just as textbooks included certain assumptions about students' bodies and minds (in chapter 1), so too did written tests. Following a previous focus on mental and bodily discipline, math anxiety began to be about written testing.

Certain historical cases emphasize changes in math testing, how they became less obviously staged though nonetheless still performative. The nineteenth century witnessed many stories of public oral examination, and the early century reports of the Troy Female Seminary, in particular, solidified gendered expectations for stage fright in math testing. The U.S. Military Academy at West Point meanwhile grew the relationship between math testing (at the blackboard) and written certifications. Emphasizing standardization, written tests in math became variously popular, and math played an important role in redefining "student performance" generally. Bringing such trends together, the creation of "math anxiety" emerged as a late twentieth-century phrase and phenomenon. In sum, stage fright and math anxiety became connected through expanded notions of testing in and with mathematics.

## Frightening Public Examination at Troy Female

Even before Vassar's "experiment," women's education faced heightened scrutiny, particularly in the tradition of public oral examination in mathematics. A regular feature of school life in the nineteenth century and earlier, "public examination" or "public exhibition" occurred at the end of completed studies, usually in the summer, and involved pupils' oral responses to questions before an audience of teachers, peers, family members, news reporters, and other community members.[12] Like N. W. T. Root's parades, drills, and gymnastics (discussed in chapter 1), such events advertised the school and its teaching, allowing students and teachers an opportunity to show off. Students' performance, often literally on a stage, also provided an excuse for the public airing of worries about their studies and their health. As more women's schools offered "higher math," their public examination inspired debates about their abilities and their potential exposure to deadly stage fright.

As women's schools' math offerings expanded, so too did worries about their unhealthy exposure. While women were not as numerate or literate as their male counterparts in colonial America, new educational programs after the Revolutionary War taught both men and women arithmetic. Some educators, such as Benjamin Rush, broke new ground by arguing that women should be able to keep household accounts and help in a family business.[13] However, the emerging ideal of femininity was opposed to market materialism, so there were many arguments against women learning about numbers, as many educational historians have recovered.[14] Algebra, geometry, and trigonometry proved even more controversial. "GEOMETRY. The sound of this word in reference to females, is very terrific," expounded James Fishburn, an educational writer, in 1828. "Parents startle at it as though it possessed some talismanic power of converting their delicate daughters into tempest-beaten rovers of the deep, and sun-burnt surveyors of the forest."[15] In focusing on the place of navigation and surveying at the end of a mathematical course of studies, Fishburn succinctly demonstrated the reasons parents found math unfit for their daughters: mathematics hardened young women's "delicate" bodies; in the guise of surveying and navigation, it exposed them to pain and disease in a distant wilderness. Resulting stories publicized the supposed results of the educational system: that sickness—even death—came from women's oral examination in mathematics.

As early as the 1820s, women's math exams drew public scrutiny. When the famous autodidact Emma Hart Willard took over the Waterford Academy near Albany, New York, her curricular choices inspired political debate.[16] As usual, the school ended its academic year with a public examination when members of the community were invited to hear the pupils answer questions on their studies, and these events became a regional spectacle when the more advanced students started to answer questions in geometry.[17] Mary Cramer, the daughter of a local politician, was the first, and the audience was variously suspicious and supportive. Some found Cramer's performance literally unbelievable. As educator Henry Fowler remembered, "Some said it was all a work of memory, for no woman ever did, or could, understand geometry."[18] In other words, some audience members thought that Cramer had been taught to mimic mathematical thought as an animal might. Others found such demonstrations inspiring. One member of the audience sent an anonymous letter to the *Albany Gazette*, later reprinted in the *Saratoga Sentinel*, that began, "On

Tuesday, for the first time in my life, I had the pleasure of hearing classes of young ladies, from ten to eighteen years of age, demonstrate with correctness and promptitude the most abstruse propositions of Euclid." The correspondent continued, "the young ladies manifested proficiency, which would astonish those . . . who yet remain incredulous in regard to the powers of the female mind."[19] For the anonymous "Albanian," Cramer and her classmates demonstrated the possibilities available in a new educational regime. As both of these responses show, women's oral exams in math faced a divided public.

After Willard became head of the school in nearby Troy, New York, she converted naysayers by soliciting powerful supporters. On visiting the Troy Female Seminary, the Revolutionary War general Gilbert du Motier, Marquis de Lafayette encountered an astounding display. The students, arranged in ranks, greeted him outside the school, where they had hung patriotic decorations. Through the scene, a committee of local matrons approached him and read a formal letter, in which they called the female students "a living testimony to the blessing conferred by that Independence, which you, sir, so essentially contributed to establish, and in which our sex enjoy a prominent share."[20] After he expressed his pleasure at the welcome, Lafayette ascended the steps of the schools to meet Willard, passing the inscription, "WE OWE OUR SCHOOLS TO FREEDOM; FREEDOM TO LA FAYETTE."[21] The letter and inscribed decorations made the message of the meeting clear, in geometrical reasoning: Lafayette's actions led to the creation of schools that offered women novel academic opportunities, including in mathematics. Few would dare criticize an institution endorsed (even implicitly) by such an influential Revolutionary War hero—though the existence of Willard's endorsement campaign indicated broader tensions.

Soon after, the highly publicized story of Lucretia Maria Davidson marred the reputation of Troy Female Seminary, when she experienced acute (and deadly) fright at her oral exams. In 1824, Davidson enrolled. She had been an eccentric child, preferring to be locked in her room, where she could write poetry and sew pieces of paper together. Her family considered the Troy Female Seminary a possible cure for her socialization, and she seemed to do well at first. She wrote to her mother that she enjoyed the classes, the other students, and even Willard herself. But Davidson developed an intense fear of public examinations. As early as two months before the exam, her letters showed signs of stress: "Oh, horrible! seven weeks, and I shall be posted up before all Troy, all the students from Schenectady, and perhaps a hundred others. What shall I do?"[22] She stayed on, though she became so sick that Willard had to send for a doctor. Davidson did pass her exams, even receiving extra praise, but her level of fright did not diminish: "Every unusual movement startles me. I am constantly afraid of something to mar it."[23] She continued to feel ill, tried to attend a different school, and ultimately died three months after the examination, on May 12, 1825. Regional reports concluded that Davidson had died from her worries about examination at a school known for its challenging math classes.

Davidson's death also inspired a bigger campaign against women's oral examination. Regional reports, after all, seemed to agree that Davidson died of overeducation. She rarely complained about her health before she began studying, but she was invariably sick afterward. Even her successes did not bring her out of her worry but seemed to heighten it instead. After 1825, Davidson's mother started to publish her poetry, along with biographical notes criticizing women's academic testing. Such demonstrations reached the height where, in 1841, Davidson's mother published selections of poetry along with a nearly one-hundred-page biography written by Catharine Maria Sedgwick. Though Sedgwick claimed not to "express or imply any censure of the 'Troy Seminary,'" she did blame the school (at least indirectly) for Davidson's death, calling the exams suicidal.[24] Davidson's biography, for Sedgwick, became a way of raising awareness about the evil effects of oral examination: "Our poor little martyr may not have suffered in vain, if her experience awakens attention to the subject."[25] Following Sedgwick, the midcentury saw an increased concern with women's testing, especially their oral examination in mathematics.

Such public campaigns made women's fright at math testing seem a widespread problem, in part because stories of women's math confidence did not appear so widely. Only many years later did a graduate of the Monticello (Illinois) Female Seminary write, "I had a natural distaste for mathematics, and my recollections of my struggles with trigonometry and conic sections are not altogether those of a conquering heroine. But my teacher told me that my mind had need of just that exact sort of discipline, and I think she was right."[26] Others, like an anonymous student from the Abbot Academy in Massachusetts, remembered creating games out of their math confidence. "I became enamored of mental arithmetic," she said, "and carried my Colburn's Sequel [textbook] back and forth from school, trying to puzzle my father and brothers over the examples I had conquered."[27] These memoirs did address some of the preconceptions of women's math worries. But because these comments were not widely circulated, the association of fright with women's math examination remained well into the nineteenth century.

Still, Davidson's story did not prevent a huge increase in women's national opportunities for math classes. Educational historian Margaret Nash surveyed ninety-one schools of the time and found that 19 percent offered algebra in the 1820s, while 67 percent did in the 1830s. Meanwhile, the percentages associated with geometry more than doubled, from 34 percent in the 1820s to 74 percent in the 1830s.[28] Furthermore, at a time of drastic regional differences, New England schools showed the greatest opportunities for young women who wanted to learn mathematics, with 90 percent offering geometry.[29] For one example, Catharine and Harriet Beecher consciously followed

Willard's lead and opened a school in Hartford, Connecticut, that listed Day's *Algebra* and Euclid's *Elements of Geometry* among its studies.[30] But their educational program was short lived, and Catharine Beecher's advocacy of arithmetic faced scorn from the community—even from her own students.[31]

In offering math for women, educators faced the problem of justifying it as appropriate, especially in light of public oral exams. As I have analyzed in more detail elsewhere, some relied on a modified rhetoric of mental discipline.[32] "It may be asked of what use will the study of Algebra, Geometry, &c. &c. be to young Ladies?" began the trustees of the new Female Department of the Providence High School.[33] They and others agreed that mental discipline provided the proper response, but they emphasized the importance of training the mind for further study in school. The Providence trustees asserted that "the mind must be *prepared* for the reception of ... knowledge,"[34] while the director of the South Carolina Female Institute used mental discipline to mean "a habit of correct reasoning, and a method of pursuing knowledge" important for further academic work.[35] Thus math trained the mind for further schooling.

Others dropped the rhetoric of mental discipline entirely, in favor of ideals about future mothers. John T. Irving, for instance, tried to choose those subjects most suited to "the domestic life of a mother" when he and his colleagues put together the curriculum for the first New York "high school for females" in 1826.[36] She should, according to Irving, be a levelheaded, affectionate companion to her husband and a knowledgeable, observant teacher to her children. Irving's system, then, included arithmetic and geometry so that the students could pass on such lessons to their children. In doing so, he tried to avoid prejudices against women's education, particularly their math education, through emphasizing "those domestic stations, for which providence has designed them"—that is, their roles as wives and mothers.[37] The domestic educator Maria Budden advocated the improvement of women's math education through recourse to the notion that women spent all their time gossiping and telling false tales. They needed geometry, for it would give them "a predilection for proofs and demonstrations" that would make them less likely to spread rumors.[38] Similar arguments reappeared in popular literature of the day and increased the allure of such programs.

Almira Hart Lincoln Phelps, Willard's sister, provided a new framework. Phelps compared women's minds to flowers or fruit trees that benefited from careful attention and pruning. "We may now consider the human mind as a garden laid out before us; he who created this garden, planted in it seeds of various faculties; these do indeed spring up of themselves, but without education, they will be stinted in their growth, choked with weeds, and never attain the strength and elevation of which they are susceptible," she told the Troy classes about mental discipline. "The skillful gardener knows that his roses require one mode of culture, his tulips another, and his geraniums another;

and that attention to one of these, will not bring forward the other. So ought the mental cultivator to understand that the germs of the various faculties should be simultaneously brought forward."[39] With an original mixture of faculty psychology, educational philosophy, and horticultural metaphors, Phelps therefore justified pedagogical choices through recourse to mental cultivation, not mental discipline. Such messages spread not just throughout the education literature of the day but also throughout the periodical press, including the popular women's journals the *American Ladies' Magazine* and the *Lady's Book*.[40]

But not all educators were convinced, even in the Northeast. Calling women the future teachers of their children, not all encouraged mathematics. The Ballston Spa (New York) Female Academy, only twenty-five miles north of Troy, distributed a circular that stated that women needed education "for [their] own rational amusement and usefulness."[41] At Ballston Spa, such claims justified the teaching of unusual classes, including advanced Latin, but it did not extend to mathematics. The seminary offered arithmetic and book-keeping but not algebra, geometry, or any other form of "higher mathematics." The principal could embrace the usefulness of "practical arithmetic" for women's understandings of money, but he found algebra and geometry simply unnecessary for women's life. Perhaps, too, he feared public attack if his female pupils were given oral exams in algebra and geometry.

Even amid changing rationales, public examination in mathematics remained a mainstay of nineteenth-century American education—for both women and men. Garnering an audience, the phenomenon solidified expectations for their math communication, in part by inspiring discussions of the place of math in their studies. Such events challenged viewers' preconceptions, in extreme cases, and encouraged the adoption of educational methods, frameworks, and technologies. As math historian Christopher Phillips has recovered, people from beyond the military academy walls could see cadets' performance at the blackboard, and audiences could judge their bodily abilities.[42] Similarly, women's public examination was a check on their health, an occasion for community members to see if students seemed tragically frightened of their advanced classes after all. Still, oral communication did not long experience a monopoly on students' examination, as written testing entered into professional certifications and the Civil War–era West Point classroom.

## Written Tests of Wartime Math at West Point

During the Civil War, the army adopted written testing as professional certification for military engineering, and the policy changed the military academy experience, ultimately foreshadowing the changes in American classrooms to come. When Lincoln and his cabinet called for the amassing of 75,000 ninety-day volunteers after the 1861 attack on Fort Sumter, far too many men subscribed. The new enlists had little military experience among them, and they distrusted military rank. In the case of engineering specifically, much confusion and annoyance could be caused by poor preparation. Consequently, the Union command was forced to ignore the engineering suggestions of anyone except for proven West Point graduates.[43] In the early Confederate army, military engineering quickly became equated with military leadership, and engineers were held up as exemplars for the rest to follow. Robert E. Lee, W. H. C. Whiting, and R. E. Rodes were all academy-trained engineers and all achieved positions of high command early in the war.[44] Furthermore, Union legislation reasserted engineering's privileged status in 1863. A new congressional action not only tried to ensure better treatment by increasing pay and rank for many of the existing officers; it also simplified promotion in the Corps of Engineers by making it a matter of written examination.[45] During the Civil War, appointment to the engineers, already a popular prospect for military academy cadets, became even more desirable, and it fueled their performance in the West Point classroom, as can be seen in the cadet diary of Paul Dahlgren, younger brother of Ulric Dahlgren.

In the wake of the U.S. Army's adoption of written testing, everyone on both sides of the war knew the name Ulric Dahlgren. A Pennsylvania native, engineer for the U.S. Navy, and then colonel for the U.S. Army, he had been part of the 1864 raid on the Confederate capital of Richmond, Virginia, but he was caught with papers indicating a plan to assassinate Confederate president Jefferson Davis. Though some claimed he was framed, Ulric was a likely assassin. The U.S. secretary of war had recognized his marksmanship at Harpers Ferry, and he had earned a distinguished record of fighting in the Battle of Chancellorsville, the Battle of Brandy Station, and the Battle of Gettysburg. Also, both the Union and Confederate militaries held engineers in high regard. Therefore, when assassination papers were discovered on him, Ulric was not only killed; his dead body was displayed as a sign of Confederate superiority. The action, known widely on both sides, understandably angered and cheered many, and Ulric, an engineer, became a symbol of the war in his dramatic death and display.[46] For Ulric's younger brother, cadet Paul of the U.S. Military Academy, life would not be the same, but his lessons in math seemed not to care. Writing about Ulric shaded into writing about math as Paul and his cohort responded to the stresses of West Point's expectations for classroom communication.

Even after the initial generation of blackboard adopters left the academy, its uses remained connected to the prestige of military engineering. Ten years younger than Charles Davies, Albert Church became the professor of mathematics after him. But where Davies had the charm of a local boy who unexpectedly helped a wartime regimen, Church represented the more usual path:

receiving admission through government networks and family connections.[47] Still, Davies and Church came together through a mutual dissatisfaction with the math education they received when they were cadets, and both openly worked toward its improvement. Under Superintendent Sylvanus Thayer, they collaborated to create new expectations for blackboard work, pushing how math could build better military officers and better engineers. Church perhaps felt even more strongly about these reforms because he really wanted to be a military engineer. But the Corps of Engineers was too popular at the time, and there were no openings at the proper rank. Church instead spent much of his twenties teaching math under Davies, with occasional forays into the work of the Artillery Corps. Once named Davies's successor, Church took greater responsibility in instituting the military academy's curricular reforms by taking over the classes, duties, and textbooks named for Davies. Continuing to push math's usefulness for the sciences and military life, Church particularly shaped the military academy's approach to geometry, borrowing Davies's popular approach to calculus and extending it to analytical geometry and descriptive geometry.[48] However, cadets spoke out more openly against him and his classes, and particularly during the Civil War, Church's classes came to represent the academy's obstinate neglect of current events.

Cadet Paul showed his initial respect for his math instructors and their classroom tech. Perhaps because he expected it to be confiscated by his teachers, Paul's diary began by emphasizing the teaching talents and military ranks of his math instructors. In October, he began by noting his appreciation for them: calling the tutor "excellent" and Professor Church "very lucid" in his explanations. At first, he included not only their names but also their ranks. "I was called on to recite soon after we had entered the room," Paul wrote on October 27, 1864, "and had for my subject the demonstration of the rule for extracting the nth root of any polynomial, which I did to the entire satisfaction of both . . . Capt. A.T. Smith, 8th. Inf., our present instr., and Professor A.W. Church, head of the Department of Mathematics."[49] As an anecdote about boardwork, the diary entry linked the instructors' facility at explaining math and their military achievements, and it therefore seemed to accept the West Point justifications for the blackboard: that the mode of oral communication made better officer-engineers (as in chapter 3).

In fact, at the military academy, there was a certain utility to pleasing the math instructors, as Paul knew. After all, since the early nineteenth century, engineering had held a privileged place in the institutional structure of the U.S. Army. While infantrymen fought on foot and artillerymen used and maintained projectile weapons, engineers performed myriad supervisory tasks: creating maps, advising units on how to proceed into battle, and planning the construction of fortifications, bridges, and a host of other structures. Their low casualty rate and relatively cushy lifestyle made engineering a desirable pursuit of many graduates of the military academy. The best in the class usually had their choice of appointment, and most would choose engineering, if there were openings.[50] Both Smith and Church, though academically motivated, had to settle for the infantry, as their students likely knew. The Civil War encouraged the expansion of engineering positions and the implementation of written testing in their certification, so success in math could lead directly to technical careers of renown.

That said, it increasingly bothered the West Point cadets that their math classes did not have clearer connections to current events. Though Paul's diary included the full military titles for Captain Smith and Professor Church, it also noted his cohort's habit of calling them irreverent nicknames. Smith, because of his physical features, was "Lanky" among the cadets, and Professor Church, perhaps because of his hygiene, was "Old Stinky." Paul at first excused these nicknames as being "familiarly designated," just among the "family" of the academy.[51] But they took on new valence in conjunction with growing animosity. While the presidential election of November 7, 1864, caused "a great deal of excitement" among the students, "Old Stinky" and "Lanky" still asked for the usual boardwork. Church became annoyed that his students were clearly distracted, yelling, "You don't study, gentlemen."[52] The war seemed a poor excuse.

In fact, according to Paul's diary, cadets did attempt to pay attention to both math classes and the latest news, particularly after the 1864 election. On November 16, a lull in current events, Paul proudly recorded how he gave a report at the blackboard about "the multiplication of quantities with negative or fractional exponents," and he "'maxed' it 'cold'" in front of fourteen classmates and Captain Smith. But his distraction returned. Three days later, he recited "the summation of series by means of his auxiliary series" in the morning and attended a reinterment ceremony for Brigadier General William R. Terrill in the afternoon. On November 29, Professor Church and Captain Smith put him "under a combined fire of questions" in the morning, and he read about the sinking of the ship "Florida" in the afternoon. December 1 saw "exciting" news "from the South," accompanied by recitation in the "superior and inferior limit of negative roots."[53] For Paul, studying happened side by side with reading periodicals, checking classmates' rumors, and learning the most recent information about American politics and battles. In fact, Paul's diary recorded how he and his classmates tried to stall their answers until they could focus. For instance, on December 1, Paul never did reply to the question about the "superior and inferior limit of negative roots," because he "managed to 'bugle' it," waiting for the sound to mark the end of class. He prepared a response for the next day and made a "fair recitation . . . probably got 2.7 [out of 3]."[54] As he recorded, the West Point cadets found ways to please their math teachers and also follow the news.

In fact, cadets performed boardwork not to please their math teachers but rather to best them. The academy's general climate of competition had been associated with math for decades. A cadet's standing in his class was determined through extras and demerits: points he received in boardwork, those he lost in misbehavior, and any he gained as rewards for unusual service. The resulting number determined not only graduation order but also choice for future appointments. At a very basic level, a number determined a cadet's entire career.[55] In their math classes, cadets wished not only to achieve the best scores but also to appear to be the absolute best, even better than their instructors. According to Paul, such a scene followed an early December decision to move algebra to the early hour of 8 a.m.: "Behold us then, arrived at 'ye section room, marching in array terrible to any outsider. 'Stinky,' his weazled face wrinkled with smiles, receives us with his customary urbanity. [Cadet] Bass reports 'all present, sir', and simultaneously we all take our seats. The next thing in the programme, Stinky got up and demonstrated Sturm's Theorem to us in his peculiar lucid and technical style, after which several members of the section tried their hand at it, but could not come up to 'Stinky.'"[56] According to Paul, the cadets as well as their professor tried to demonstrate a proof at the blackboard. Though admitting Church's talent, Paul also displayed his own ambivalence toward him by using the mocking nickname. As with the system of demerits, the sense of competition also had to do with the career aspirations of everyone involved. Perhaps cadets thought if they showed themselves to be better than their teachers at boardwork, they could become the engineers their instructors aspired to be.

Despite the incentives for noticeable success in math, news of the war continued to be a distraction, especially for Paul. He followed news of Sherman's March in December, just managing to find enough time to prepare for his algebra exams after the New Year. In January, he glanced over the Books of Euclid for his new geometry class and meanwhile read the newspapers voraciously. In his diary, Paul recorded the latest intelligence about military losses and victories, noting, for instance, the Union forces' unsuccessful attempts to take Wilmington, North Carolina. He also recorded his boardwork topics and possible scores: propositions from Euclid-inspired books of geometry and numerical values, "about 2.9 [out of 3]," "about 2.5 [out of 3]." Nearing the end of the abbreviated geometry class, he joked about the two interests coming together, saying, "made a good recitation on a prop. in the 8th Book-Surf. gener. by revolv. a reg. semi-poly. &c. &c. 'Stinky' was present & let fly a few mathematical bolts of war."[57] Modifying a quote from Shakespeare's *Julius Caesar*, he indicated how the geometry classroom became a battlefield. At the other end of their instructors' "bolts," students regularly took to the blackboard stage.

Paul's performance in math depended on his worries about his family's fate in the war. He noticed articles about naval actions because his father was the rear admiral in charge of the South Atlantic Blockading Squadron. Whenever friends of his father died throughout January and February, Paul recorded their names in his diary, and he sometimes wondered about the fate of people he knew. He also read articles with special attention if they mentioned attempts to take Confederate cities because of the infamous death of his brother. On the anniversary of Ulric's public execution, Paul wrote a eulogy: "A year ago fell one of America's noblest sons & whose memory forever lives—Ulric Dahlgren. Thus perished at the age of 21 as noble a specimen of an American soldier as ever God created. His name will be handed down from generation to generation; time only serves to make his deeds brighter, & in years to come may even the humblest have it in their power to contribute their mite to raising a monument to his memory. God grant that it may be so."[58] Though Paul's specific circumstances were particularly connected to the raging war, his classmates had their own. After all, admission to the academy often followed from family connections to American politics and/or the military. Many cadets had similar stories, whether preserved or not. They similarly became distracted from their mathematical studies in the process. Church, after all, did not pick out Paul specifically when he chastised the class about their inattention; he addressed the whole group.

Given the climate, the cadets especially celebrated when outside circumstances forced their math instructors to pay attention to the war. In early March, Captain "Lanky" Smith received a reassignment to the front lines, requiring him to give up his mathematical duties and take on martial ones. Paul wrote in his diary on March 4, "A report circulated after supper that 'Lanky' (Capt. Smith, 8th Inf.) would be relieved to be Colonel of a colored regiment out West. The joy of all classes knew no bounds & several processions of first-class men went around barracks, hooting and kicking up."[59] The big party continued, somewhat surreptitiously, for at least ten days. Still on March 14, bands played music in the barracks, and cadets sneaked to the festivities, occasionally bringing contraband food or drink with them. Smith could no longer remain in the relative isolation of math studies.

Tensions increased when Paul's class passed on to study descriptive geometry, a subject that increasingly relied on writing. A somewhat unusual offering, the subject indicated the academy's joint status as scientific and military. It used geometry to motivate considerations of accurate drawings: projections, "shades and shadows," and mathematical studies of perspective. But aesthetics did not inspire Church to offer the subject at the military academy. As he explained in his textbook's preface, the subject was important not only for a student of math but also for a "civil and military engineer."[60] It was all about "representing by drawings" and developing accurate technical diagrams given a "point of sight."[61] Yes, he concluded, there was a certain aesthetic satisfaction in creating drawings that "present to the eye of the observer a very good representation of the objects." But there was also an employable skill here: the system of drawing was, for Church, of "great use to the machinist and builder."[62] Church therefore suggested how the mathematical system made possible future construction and also battlefield reconnaissance. It emphasized abilities with a pen and paper, abilities in writing.

The new focus on writing paradoxically heightened cadets' sense of stage fright, as suggested by an insert into Paul's diary (see figure 6). Titled "'Descriptive on the Brain': Or a Night in Cadet Barracks, West Point," it used the techniques of accurate drawing to produce a fantastical scene of mathematical devils surrounding a cadet. In the doodle, some of these supernatural minions play with geometrical objects: lines, cones, circles, a compass, and a mathematical model. But they mainly fix on distracting the central figure: a young man sitting at a table, reading a book, smoking a pipe, and wearing his uniform. Though the tiny devils are facing him and surrounding him (with one even on his back), the stationary cadet ignores them, keeping his head turned firmly to his studies. On the reverse of the drawing, the caption explains,

The Evil effects of studying Descriptive Geometry as shown at U.S.M.A.

For further information inquire of Professor A.E. Church, Mathematical Department.[63]

These captions reframe the scene. Though the main "Evil" might at first seem the devils misbehaving with mathematical objects, Paul indicated otherwise. The scene did not show some external hell but an internal one: math studies infecting "the brain." Without the demons, without the mathematical models, it was a scene of a studying cadet surrounded by a lot of open space. Sometimes described as "evil" or "devils" "on the brain," fright therefore was the main topic of the drawing.[64]

Despite cadet worries, connections to military engineering continued to motivate experiences in mathematics. The cadets heard about General Ulysses S. Grant's victories in Virginia, and they saw the announcement of the Confederate surrender at Appomattox Court House, where one West Point graduate surrendered his forces to another. Despite the end of the war, the academy still ranked cadets by their numerical scores, combining their academic performance with matters of social behavior and military order, and these scores depended a lot on mathematics. Not only did the system inspire cadets to calculate guesses about their own totals, but it also weighed heavily toward the required math classes, since cadets received three points per

FIG. 6 "Descriptive on the Brain" insert, in Paul Dahlgren's diary, *Private Journal. No. 5* (1864–1865). Special Collections and Archives Division, U.S. Military Academy.

recitation, and the math recitations (unlike many others) occurred six days per week for months on end. Paul cared a lot about how many times he "made a max" in math, and he proudly recorded the twenty-three times in his diary. Furthermore, most cadets wished to please their math instructors because they considered their ideal appointment to be a mathematical one, with the Corps of Engineers. By the late spring and summer, even Paul focused on engineering goals instead of news reports. When he wrote in his diary that he had read about "Booth the assassin" being "shot down like a dog," he did not worry that his math classes ignored the occasion. Thinking about Booth's death just enough to write in his diary about it, he spent "most of the afternoon on an engineering problem" on the side.[65] In Church's descriptive geometry and beyond, he and his classmates chose to do math quite often and of their own accord because of its connections to their career prospects.

In sum, though the Civil War did inspire math worries for West Point cadets, they did not make the situation public. My own analyses have been heavily weighted toward Paul's diary because there he expressed his feelings about math, as well as his sense of his classmates' and instructors'. The military academy's social mores prevented these conversations from being a news feature by themselves. Unlike at the Troy Female Seminary, no story broke about cadet antics or cadet frights. Professor Church did not make public statements about the ways the war distracted his students from their boardwork. Nor was there a public mention of cadet partying in the wake of Smith's dangerous assignment. Such gripes remained internal: internal to the institution and usually internal to the individual. As the U.S. Army's written tests expanded to analogous educational initiatives on the state level (on the Union side) and then nationally, only their public goals remained: not to make math frightening but instead to qualify young men for advancement.

## Postwar Debates about Written Testing and Math

After the Union's Corps of Engineers instituted written exams in 1863, in an attempt to make appointment and promotion more desirable, written testing quickly became considered a standard method for educational qualification. The "Regents of the University of the State of New York" followed the next year, announcing that they would offer written tests in arithmetic, geography, grammar, and spelling as a way of certifying students' readiness for entrance to regents-sponsored schools.[66] As noted by math historians Peggy Kidwell, Amy Ackerberg-Hastings, and David Roberts, written testing provided a way to standardize exams for both private academies and the academic departments of the state's public schools, and it also provided a measure for the state to use when they determined funding (more funding for the schools with more qualified students).[67] Between the Civil War and the First World War, many other professions and many other states attempted to follow these examples. As these educational movements reached national audiences, the worries of students—and educators too—came to be about the value of written testing and mathematics.

In part, math came under attack because of its uses in providing the overall framework. In both accreditation movements in professional organizations and also in schools, written exams were considered valuable because they exposed certain statistical patterns—in other words, because they were ultimately mathematical. Not only tests' implementation and analysis but also their creation needed to consider how to make them lead to verifiable quantitative information. Educational researchers from the 1860s to the 1910s therefore built careers from the creation and modification of written tests. Even when not about the subject matter of math, those tests' construction proved to be eminently mathematical. Unlike math's other classroom technologies, as Kidwell, Ackerberg-Hastings, and Roberts point out, written tests have used math in the service of mathematics.[68]

There were two other reasons for these math debates. First, as educational standardization increasingly looked at high school curricula, math seemed to be the only commonality amid a huge diversity of offerings. According to Harvard president Charles W. Eliot's 1892 survey of subject offerings at forty American high schools, college preparatory schools (public or private) offered little more than Latin, Greek, and mathematics. Students at the Phillips Academy, for instance, spent much of their class time in Latin instruction, followed by Greek and math, with some German, French, chemistry, physics, and history.[69] Latin high schools usually dropped a modern language in favor of certain sciences. Students at the Boston Public Latin School took Latin, Greek, mathematics, French, English, physics, geography, natural history, and physical training.[70] Lastly, English-scientific high schools prepared a student for industry work or perhaps a college modeled on West Point. Students at the Battle Creek High School in Michigan spent most of their time in English composition and in a full complement of sciences (physics, astronomy, chemistry, geography, natural history, and math), as well as some Latin, Greek, German, French, rhetoric, history, civil government, political economy, and bookkeeping.[71] Particularly because bookkeeping was considered a form of advanced arithmetic, math appeared prominently in any option.

A second reason for math debates in written testing remained: the new American communities of mathematicians could not agree on what math meant. Recall that both astronomer Mary Whitney and mathematician Achsah Ely represented Vassar in the early years of the New York Mathematical Society (NYMS). But American astronomers left the organization en masse on the eve of the creation of the American Mathematical Society.[72] On the first day of the Congress of Mathematics and Astronomy at the Chicago World's Fair of 1893, the members voted to break the meeting along new disciplinary lines: in one room, the Mathematics Congress and, in another, the Astronomy Congress.[73] They could not even talk to each other, because they disagreed so fundamentally about why they were meeting.

The schism in Chicago followed years of debates concerning the proper meaning of mathematics. A style of graduate education had been developing that encouraged an appreciation of math divorced from real-world applications, leading the student to be, in the words of G. B. Halsted, "lost in the highest cloud-lands of mathematics."[74] Contrary to the Neoplatonist approach, astronomers emphasized detailed, real-world examples as key to proper pedagogy. The astronomer T. H. Safford became a member of the NYMS with Vassar's Ely and Whitney, yet he became a spokesperson for math reform along the lines of new, applied programs.[75] Much of his own training had been as a "calculator" and then observatory director at Harvard, Chicago, and Williams, so he hoped to see math reflect the importance of calculation tools and quantitative shortcuts in the physical sciences.[76] The fundamental disagreement appeared in the initial discussions of nationwide standardization—and helped fuel the debates about written testing in mathematics.

Even before the first nationwide tests were proposed, there was reason to worry about mathematics. Though variously theoretical or applied, it appeared in every curriculum for every student in every school. In an era of increasing diversity, it therefore seemed the best place to test the idea that there could be uniform qualifying exams in education nationwide. Furthermore, America's standardization movement already urged the values of quantification. Numbers, when written, could be combined, coordinated, and ultimately circulated easily.[77] Educational researchers used these trends to propose a new understanding of "performance": a "measurable [esp., quantifiable] action" especially in the classroom.[78] Expanding performance made possible a performance anxiety beyond stage fright, one that could encompass technologies of oral communication and written communication, from the blackboard to the test.

## Redefining Student Performance in Math and Beyond

In math, especially, written tests distressingly expanded the consequences of student performance. Previously, boardwork was a major way of determining how well a student understood the material. It provided marks that could even lead to promotion and appointment, as at the military academy. But it was also tied to individual instructors and individual institutions. Written testing heightened the stakes, particularly in its increasing application to whole regions and nations. For various professions, it could be a mechanism for denying credentials for those who could not pass licensing exams—or for those who could not take them because of their race, ethnicity, class, or gender. For education, it could be a way of enforcing uniform curricula amid increasingly diverse students and schools. As the New York State Regents began showing in the 1880s, it could be a way of determining school funding, if not enough students seemed to understand the material after relevant classes. It could also confer new qualifications, such as a "college ready" diploma from a Regents-sponsored school.[79] Such movements promised to make more opportunities accessible to more people, as math historian Theodore Porter has argued, but as educational historian Nancy Beadie has countered, they also discouraged participation.[80] In response, a new crop of educational researchers took up the project of nationwide written testing, and they articulated a new meaning of "student performance" grounded in mathematics.

Math education continued to be debated after the Chicago schism of the American Mathematical Society in part because of the Committee of Ten's report. The "Committee of Ten" brought together one hundred educators to explore the standardization of high school curricula and the instituting of written tests around nine subject areas.[81] After initial publication, responses from others were mainly critical. The criticisms leveled against the math committee matched that overall reception—for example, that the recommendations were not practical, because actual teachers were in the minority of the committees. At the 1894 National Education Association Convention, Superintendent J. M. Greenwood of the Kansas City High School used the argument to speak out against object lessons in arithmetic: "To the Committee of Ten, and to the Committee of Ninety [i.e. the subject specialists], I will say, that the only way a boy can learn arithmetic is to study arithmetic and not to mix it up with other things." The math committee had framed some of these recommendations as time-saving initiatives, but Greenwood warned that the strategy was not tested yet. In fact, he chastised the math "specialists" for not doing the testing long ago: "Let them take their own medicine first!"[82] Similar worries at first led to an uptick in American publishing about education; dozens of articles in educational and popular periodicals reviewed the report and found it lacking.

These criticisms quickly turned to educational research programs, as the report became a useful point of comparison for anything new. The Boston superintendent of schools reflected, "By general consent already, it would seem, has the Report been accepted as a convenient standard of reference in discussion. Across the chart of our educational theory and practice there has been drawn, so to speak, a meridian line, by noting his departure from which one may easily define his educational position. In this respect, undoubtedly, the value of the Report will be admitted by those who are least inclined to accept its recommendations."[83] The math committee's recommendations were similarly a "meridian line" for future work, even from the most vocal critics. The day after Superintendent Greenwood yelled about the content of the math committee's suggestions, he started advocating a high school elective system that gave students greater control.[84] He explicitly framed these views in response to the Committee of Ten's report, and in doing so, he admitted that some merit could be contained in their suggestions, if only as a point of comparison.

In part because of such responses, many teachers, psychologists, professors, and others used written tests in arithmetic as a way to advance a specific field: what came to be called "educational research." At first observing students and advising superintendents on the teaching in their schools, the psychologist Joseph Mayer Rice began surveying arithmetic students circa 1900. As part of his project of figuring out how to get "the ship of pedagogy" from being "waterlogged in a sea of opinions," he distributed six thousand tests throughout the country: eight word problems for grades 4–8.[85] Though he later realized he could not be sure what was actually covered throughout the country, he made a guess and found some surprising results. Compared to his earlier written tests of spelling, there was more variation among schools, especially in grades 7–8. Also, surprisingly, the statistical patterns did not correlate to ethnicity or measures of students' home life; it did not relate to individual teachers or even the amount of time those teachers chose to devote to the subject. Instead, it seemed the test scores grouped by city: some cities did very well and some did very poorly. He concluded that superintendents played the biggest role in student performance, reinforcing the need for his own practice.[86] In his study and beyond, Rice made it clear that he and his colleagues were not just contributing to "the sea of opinion" surrounding nationwide education. Instead, they were the experts, the specialists whom superintendents needed to consult so that they did not become "waterlogged." In short, through testing, observing, advising, and especially quantifying, educational psychologists' ideas rose to the top.

From Rice, the project of developing broadscale written tests—especially in arithmetic—built new understandings of student performance. Edward Thorndike, of Columbia's Teachers College, attracted many graduate students while literally redefining key terms in the field. His 1898 article about intelligence testing in animals framed animal behavior as "intelligent performance," investigating what could be observed and quantified as being related to intelligence and not just researcher interpretation.[87] His 1903 textbook *Educational Psychology* and related work extended "performance" work from animal behavior to student observation and testing. Thorndike's work in "student performance" strongly advocated the quantitative measurement of students, as it brought techniques from animal behavior to understand student psychology.[88] Thorndike used such perspectives to modify educational orthodoxy, as in his criticisms of mental discipline, showing that Latin did not lead to better learning in other areas—and neither did mathematics.[89] In building a new sense of performance, Thorndike and his graduate students worked on a wide range of written arithmetic tests in order to gauge student learning. The superintendent for Albion, Indiana, schools, while working as Thorndike's student, administered seventy-seven written tests in addition, multiplication, fractions, and word problems, finding better scores from girls than boys. Another Thorndike student, a Wisconsin teacher, administered arithmetic tests to three thousand students spread among six states, finding that arithmetic built certain quantifiable, observable student abilities but not others.[90] The research fleshed out the idea that mathematical study was the best way to evaluate math students, strongly encouraging written testing over oral examination along the way.

The quantitative nature of these studies led to the tests' popularity and adoption among teachers. Teachers from Chicago and Detroit called the reports from Thorndike's students "scientific" and "rational," a way of "beginning to standardize" and estimating "the influence exerted by any method, material, or teacher."[91] In other words, "student performance"—in the new meaning from the field of educational psychology—seemed the best way to extend the reach of both learning and teaching in America. Still, teachers did modify the tests to better fit their needs. Though the tests from Thorndike and his students did have to be timed, the Detroit version took inspiration from the tradition of N. W. T. Root, who advocated arithmetic drills. Unlike the earlier oral exams, the new "speed tests" were written on preprinted paper, directing students to do as many questions as they could in addition, subtraction, multiplication, division, the copying of numbers, and the translation of word problems (from words to numbers). Marketed to the country (and the world) from the teacher Stuart Courtis, speed tests became widely popular as a quick way of determining the arithmetic performance of entire schools. From the Liggett School for Girls, these exams spread to the Detroit Public Schools and famously became of measure of "school efficiency" in Boston, Cleveland, and New York City. In a year, four hundred thousand of these written tests were administered in forty-two states, and during the product life of the "Courtis Speed Tests" (1909–1938), over twenty million sold throughout the world.[92] Not yet a nationwide standard, these tests "efficiently" encouraged student evaluation to shift from the blackboard to the paper.

The early adoption of arithmetic tests emphasized them as sources of teacher inspiration, not student worry. Shortly after Courtis began administering speed tests to his own students, he wrote to the school yearbook about his feelings toward what he called the "strange exercise": "A hundred heads bent over a hundred scurrying pencils, a silence vibrant with the energy of a hundred minds at work upon a single task; a signal [to finish], a pause, and a sigh of relaxation from a hundred lips as from the lips of one."[93] Clearly romanticizing the tests, Courtis's description was tellingly not about mathematics. It was about the collective body: "a hundred heads . . . a hundred minds . . . a hundred lips" moving with "a single task," moving as "one." Moreover, Courtis waxed eloquent about how these students were engaged in writing. Noting the presence of "a hundred scurrying pencils," he still focused on a "vibrant" "silence"—these speed tests were in contrast with the oral public examination. His romanticizing of collective writing partially inspired Courtis to keep refining his tests and then marketing to them to bigger and bigger audiences.

Still, these written tests did inspire worries in the teachers, though not yet about student reactions. Courtis revised to try to separate out students' abilities. He found his tests of quantitative "reasoning" had more to do with reading than math knowledge and that all arithmetic questions did not scale from simple to complex. Mainly, scores fluctuated too much to correlate with an individual's strengths and weaknesses, and Courtis therefore marketed speed tests with the slogan "Measure the efficiency of the entire school, not the individual ability of the few."[94] In response, self-proclaimed educational researchers developed written tests that they hoped would rank arithmetic knowledge according to improvement—the sort of tool that emphasized "ability" over broadscale "efficiency" instead. Marketed as "scales," "fundamentals," and "diagnostics," these tests directed students to write on preprinted paper, answering as many questions as they could in the allotted time. Whether framed as difficulty, practice, or habits of mind, researchers worried about the tests' validity (whether they measured what they were supposed to measure) but rarely the effects on students.[95] Expanding the reach of arithmetic tests in the 1910s and the reach of algebra and geometry tests in the 1920s, these projects encouraged evaluating written tests on preprinted paper as opposed to oral boardwork so that evaluations of student performance could be scaled up.

When educational researchers came to worry about students, it was about the use of such tests in limiting their opportunities. After World War I, tests came to follow a model of the U.S. Army's Sanitary Corps, which aimed to weed out those "unfit" for service through measures of intelligence. Thorndike and his graduate students, following the army work, devised tests that were supposed to show who should not be required to take mathematics. After all, educational researchers had proposed a drastic revision of mental discipline, so math was no longer necessary exercise for everyone's minds. Furthermore, the eugenics movements had partnered with educational policymakers to advocate (highly suspect) claims that "mentally unfit" students wasted taxpayer money. Generally ignoring the racism, ethnocentrism, and class biases associated with these arguments, educational researchers nonetheless worried that such testing would deprive some students of their opportunities.[96] One Thorndike student explained how she felt "in a democracy . . . no child of normal intelligence should be deprived of the opportunity of becoming acquainted with a realm of thought, which has meant so much for the advancement of science and of civilization."[97] The talk of "normal intelligence," of course, assumed that some students could not "profit by it," reifying the overall framework. Others criticized such assumptions outright. Mathematician David Eugene Smith wrote that such a test was "at its best" more efficient than previous options but "at its worst it leads into a determinism that is more dangerous than the extreme form of Calvinism which left each individual without hope."[98] In other words, when the stakes were as high as access to certain subjects, it was equivalent to arbitrarily determining religious salvation—by written, preprinted tests, no less.

Meanwhile, the Committee of Ten's successors similarly redefined the stakes of written testing nationwide. Between 1895 and 1901, the Kentucky Educational Association, the Missouri State Teachers' Association, and the Mississippi State Teachers' Association all formed their own committees with their own subject specialists to determine the best methods of teaching at secondary schools in their own states.[99] On the national level, a general question raised by the Committee of Ten inspired the creation of the College Entrance Examination Board. Could there be a standardized (written) exam that would provide a certificate saying that school students had the proper level of knowledge to attend any one of a list of colleges and universities?[100] President Eliot of Harvard and Nicholas Murray Butler—once an organizer of the Burial of the Ancient (see chapter 4) and now Thorndike's colleague—pushed the idea of teaming up with other universities in setting standards for admission.[101] In 1899, at the annual meeting of the Association of Colleges and Preparatory Schools of the Middle States and Maryland, Butler read a paper advocating uniform college entrance examinations "with a common board of examiners," and Eliot publicly showed his support.[102] Throughout the next calendar year, Butler amassed a board of representatives from colleges and secondary schools, very similar to the composition of the Committee of Ten and its subject conferences. In 1900, the College Board was founded, with members from Barnard College, Bryn Mawr College, Columbia University, Cornell University, Johns Hopkins University, New York University, Rutgers College, Swarthmore College, Union College, the University of Pennsylvania, and Vassar College.[103] Despite some geographical spread, the College Board's initial exams of nine subjects mainly attracted hopefuls for Columbia. Still, from 1901 to the 1920s, the tests eventually reached much of the nation and even led one of Thorndike's graduate students (at Columbia) to worry about the scoring of such a large number.[104] After the work of many educational researchers, written testing in math finally replaced oral examination as the major marker of college readiness.

Stopping here, it would be helpful to gauge actual student reactions to the work of educational researchers—these changing understandings of written testing and student performance associated with math—but they generally do not survive. Perhaps tests were repositories for student doodling—something along the lines of the insert "Descriptive on the Brain." The Archives of the History of American Psychology has a cache of tests, as do other archives around the country. However, as they have prioritized information about the tests' construction and not their reception, blank Courtis speed tests have been saved, not ones that have been filled out.[105] Further research needs to be done on students' reactions when faced with the new ideals of student performance in and with mathematics.

In sum, educational researchers certainly controlled the message. In newspaper articles, specialist journals, conference proceedings, and school boards, they claimed to speak for superintendents, teachers, and others. Filled-out student tests and other ephemera have generally not been saved because of their success—the legacy of the idea that educational researchers' initiatives were the main ones that mattered. Still, it must be admitted that educational researchers certainly raised awareness about American education. They documented disparities among schools, regions, and classrooms, and they advocated for equal opportunities. On the whole, they worried about issues of fairness and validity, and they tried to make changes when their neglect was pointed out to them. Still, their voices did prevent others' from being heard, and their test questions perhaps did provide implicit advantages for those already privileged in American classrooms of their day.[106] In their zeal to administer written tests, educational researchers literally redefined performance and encouraged its application to students, and that, in particular, has had lasting effects in American classrooms and beyond, especially in terms of math anxiety.

## Math Anxiety, Performance, and Stage Fright

The creation of *math anxiety* as a term and a phenomenon has been connected ultimately to these movements of written testing in and with mathematics. Since the 1970s, educational researchers have shifted focus to students' negative feelings about math and/or about their abilities to do mathematics. The phenomenon has been consistently called a "disease," though its name has changed in response to changing medical terminology. In the 1950s, it first emerged as a "phobia" surrounding math, what math teacher Sister Mary Fides Gough called *mathemaphobia*, which she suspected could explain many cases of students (especially girls) failing in their classes from sixth grade to college. Writing for nonspecialists as well as specialists, Gough hoped to draw attention to medical/psychological roots of math failure so that there could be a "cure."[107] Since the 1970s, there have been many attempts to use various scientific methods to demonstrate the existence of such a disease, while its name has gradually shifted to "mathematical anxiety" or just "math anxiety." Regardless of the specific name and even the specific "cures" advocated, these techniques and research studies have come together through their focus on student performance and the management of written testing.

Student performance has been literally written into the definition of math anxiety. Psychologist Mark Ashcraft has defined it as "a feeling of tension, apprehension, or fear that interferes with math performance."[108] Like many other psychological diagnoses, for Ashcraft, math anxiety has to do with unproductive feelings, those that interrupt the usual measures of student behavior in math—the sorts of things that Gough characterized earlier as leading to "failure." But that failure, both Ashcraft and Gough have agreed, should not be interpreted as some inability to answer questions and do calculations but rather as the interference of physiological and behavioral factors that prevent productive action. Following Thorndike's foundational term, math anxiety is a matter of (math) student performance.

Extending the connection with behavioral performance, educational researchers have spent much of the twentieth century (and beyond) identifying the quantifiable, measurable aspects of math anxiety. In the 1970s, psychologists Frank Richardson and Richard Suinn proposed a "mathematics anxiety rating scale" (often known as MARS). Interested more broadly in counseling, they developed a psychometric scale to show the condition existed while proposing how it was associated with test anxiety and could be managed with behavioral therapy.[109] As with the developers of speed tests, Richardson and Suinn revised their own scale and encouraged others to do so too. Though different scales have arisen for measuring math avoidance (as opposed to anxiety) or "mathematics attitude," MARS remains the standard in counseling, leading to a slew of recommendations for the classroom, center, and doctor's office. The link to measurement has gone further as well. Reflecting the changing techniques of psychology, Sian Beilock revisited some of the initial claims of math phobia/anxiety using brain imaging in the early 2010s to reassert that sufferers are not unable to do math but register a form of pain when encountering or anticipating situations involving mathematics.[110] The shift extends the measurement of performance (as productive behavior) to something approaching the measurement of "stage fright" (as unproductive behavior).

Math anxiety also has emerged as a component of math classes' reliance on explicitly timed, implicitly written, high-stakes tests. In 1990, mathematician Ray Hembree reviewed all the published literature on math anxiety, finding that it was associated with negative attitudes concerning math and poor marks on relevant achievement tests.[111] MARS had assumed the relation between math anxiety and test anxiety as early as the 1970s, and publishing about both has emerged as a consistent trend in the work of Hembree as well as (more recently) Ashcraft and Beilock. In more recent workshops and some articles, teacher Gary Scarpello has talked about American education's reliance on timed, high-stakes testing from grade 4 to college as a reason for the prevalence of math anxiety.[112] Though none of these researchers have looked to an explicit link with writing, such situations do involve written communication, as opposed to the earlier model of boardwork, and many of the underlying assumptions (such as "student performance") come from the early projects of broadscale written testing.

The treatments for math anxiety have extended to appreciating variety in technical communication, a main focus of the conclusion. Anticipating Hembree's work and in response to it, the National Council of Teachers of Mathematics has synthesized recommended strategies, and at the top of their list, they note the importance of "accommodating for different learning styles" and "creating a variety of testing environments."[113] In other words, math teachers should not rely on drills or achievement tests, because such evaluations only work for certain students. Teacher Marilyn Curtain-Phillips, among others, has translated these broad statements into specific classroom activities, including lessons with visualizations, computers, and even specific techniques that have to do with performance and theater.[114] Though the stage of the math classroom is no longer used so centrally in evaluating students (given a move away from public oral examination), work on math anxiety still draws analogies from stage fright in Curtain-Phillips's work and beyond. Math anxiety and stage fright have been connected through written testing in and with math, ultimately exposing the performative dimensions of math communication from the nineteenth century to today.

# Conclusion

• • • • • • • • • • • • • • • • • • • • • •

## Math Communication from
## STEM to STEAM

In following the show of learning math, this book has considered how math communication has begun with reading; how math communication has been practiced in prohibited ways; how math anxiety has emerged from classroom tech; how math communication has been (literally) theatrical; and how math anxiety has come to be about written testing. But there are still a few questions remaining: How have realizations about math as performative occurred, and what difference could they make for both math people and performance people? How do the historical experiences documented here speak about math experiences in general; What have these stories told us about why math has been so hated, avoided, and worried over in the nineteenth century and since? Finally, what do they say about math anxiety in general and possible ways to cope with it?

The first question can be answered through recourse to the STEAM movement. STEAM relies on the framework of STEM (science, technology, engineering, and mathematics), which occasionally has been approached as a "delusion," in the words of Andrew Hacker: a monolithic educational regime that promises to erase anything more human and more humanitarian. Hacker's stance has been developed mainly in response to the more popular perspective that STEM-centered education can provide solutions to many of our generation's problems, including decaying infrastructure, changing climates, and increasing food insecurity.[1]

STEAM recognizes the value of STEM while also arguing for the importance of its integration with the arts. Some educators and journalists have claimed that the arts can make STEM better in an instrumental way by providing new perspectives and skills that will make better products in the lab and workshop.[2] Others, especially scholars of psychology and mindfulness, have recognized that we cannot know what will be useful in the future and therefore the arts should be pursued for their own sake, as STEM fields often are.[3] Rarely, though occasionally, academics and journalists have asserted the perspective that the arts can act as a gatekeeper for STEM work, allowing for critical understandings of ethical, moral, social, and cultural implications.[4] Responding to these perspectives, this book has been a study of math communication through the frame of performance, expanding STEAM by providing avenues to both math appreciation and performance appreciation. After all, the best of the STEAM movement, it seems, does not merely provide add-ons for STEM but articulates benefits for both STEM people and art people. Urging us to recognize how studying and communicating about math involves a considerable amount of theatrical performance, this book adds to both math and performance.

Principally, this book urges recognition that American math has been performative. Shortly after the Revolutionary War, math classrooms were sites of public examination, competitive drills, and even half-naked parading. These activities relied on the political power of showing off, convincing stakeholders of the successes of early schools and colleges. Students, especially at the college level, took the convention and made it their own. Through silly funerals for their math books, they practiced the ways that they were encouraged to show off, creating their own public performances with costumes, stagings, blockings, songs, and props. Theirs was performance art with an academic edge, a tradition that allowed audiences to understand the performative dimensions of institutionally sanctioned displays of mathematical strength.

Even drastic changes to math's communication conventions did not remove these links with performance. As the introduction of blackboards emphasized certain (military-inspired) rules of oral communication, the architecture of the typical classroom changed, gaining not only painted/slate walls but also raised platforms. The expectations of boardwork, occurring in conjunction with these changes in classroom tech, invited students regularly to take the literal stage. Instances of student rebellion followed, communicating collective worries about the changes to communication conventions, and so too did actual instances of student-generated math plays. Finally, even with the rise of written testing, theater did not disappear as a point of concern. Stage fright, under the guise of "math anxiety," continued to haunt students, now at the preprinted exam as well as the blackboard. Similarly, the new expectations of written exams relied on changing notions of performance, especially from educational researchers who looked for "student performance" as something measurable and quantifiable in itself. Well into the twentieth century and beyond, math and performance remained intertwined.

The STEAM frame has a major benefit for the arts as well as for STEM. It shows how the historical math classroom can be a site for interesting performance studies. Some of my earlier work has begun to be used, for instance, as groundwork for a chapter in *Rethinking Campus Life* about the history of drag shows, written by educational historians Margaret Nash, Danielle Mireles, and Amanda Scott-Williams. Partially inspired by Clyde Fitch, who performed as Mattie Mattix at Amherst before having a Broadway career, they have synthesized articles from both math and theater, emphasizing how these historical traditions should be seen as combined, especially in studies of higher education.[5] Such articles begin to indicate how theater scholars can integrate math classrooms/events into their histories of performance practices.

Also, recreating math funerals and plays (in part or in full) can provide new possibilities for productions. Such events have been incorporated into alumni reunions, notably at Yale's 2001 tercentennial, which included a Burial of Euclid procession among the festivities.[6] The Yale recreation also followed the tradition of alumni reminiscences, such as Hamilton's Half-Century Annalist Letters, which emphasized quirky student traditions that asserted class and college pride. More recently, my performance of songs from *The Mathematikado* at the 2017 British Science Festival provided an opportunity for collaboration among historians, musicians, and actors. Within the frame of a science outreach convention, arguably the longest-running one in Europe, *The Mathematikado* event was nonetheless advertised as a "theatre event" from the University of Brighton as well as a "maths" event from the London Mathematical Society.[7] Interdisciplinary advertising can result in wide appeal.

The recognition of math's role in STEAM allows math practitioners and researchers to rethink their work as well; some mathematicians already have attempted to claim that their work is more artistic than scientific. American mathematician Jerry King, citing British analyst G. H. Hardy, has argued that beauty is the most important factor for mathematical research and has promoted its aesthetic considerations along the lines of the arts.[8] Many others have conceded that math is ultimately judged on artistic grounds, arguing that it is not like the sciences, because its use and appeal cannot be fully articulated.[9] As in the extensions of these classic arguments, math's relationship with performance can inspire conversations about both mathematical tools used in performance practices as well as the benefits of variously categorizing math teaching and research. When we recognize math as performative, in short, we can be more receptive to its artistic qualities.

Also, understanding math history through the arts allows for further developments in math appreciation. As topologist John McCleary and philosopher Audrey McKinney argued in response to Thomas Kuhn's *Structure of Scientific Revolutions*, it seems math changes differently from the sciences.[10] The sciences assume a framework through which theories are viewed, but when there are enough observations that do not fit the framework, then the framework has to change completely. (A usual example is the shift from Newtonian to non-Newtonian mechanics—even the terminology emphasizes the sense of total replacement.[11]) While math can be seen as asserting a similar framework for viewing the relevant world, its changes are slow to come and somewhat partial by comparison. In their article, McCleary and McKinney therefore suggest *reorientation* as a more appropriate term for math history than "revolution."[12] Their conclusion has further implications when considered in conjunction with their further articles that the history of math might better be classed under art history rather than the history of science. Not only does it suggest that art scholars, too, might benefit from the idea of reorientations, but it also begins to explain why math imperfectly fits models of change in science. Appreciating all these perspectives, it seems clear that some mathematicians would criticize the place of math in STEM but support the place of math in STEAM.

One striking aspect of these STEAM explorations is the cultural power of math relative to performance today. Usually today, math is understood to be a component of scientific and technological careers, and its teaching (and some research) often receives support from politicians who argue for STEM, that we should solve problems by making and doing.[13] That is not to ignore the bad feelings toward math, the ways that it has been misunderstood and even "hated." Among signs of its cultural importance, in fact, are such statements. It has been debated for over two hundred years because it has been surprisingly common as a school requirement, a component of a written test, an introduction to boardwork, a gatekeeper for career development, and a site of anxiety. Studies of performance, particularly theater, have not been required so pervasively.[14]

Have historical actors noticed the performative component of mathematics? Yes and no.

The presence of the Burial of Euclid—and its amazing prevalence throughout the nation—introduced thousands of students to the work of organizing a public spectacle of mathematical achievement, as reviewed in chapter 2. Their collective *exercises* (to use their term) of parading and processing indicated that they understood some connections to broader performance traditions. Though we do not have reminiscences from the majority of participants, some students did recognize how these components combined math and theater. (The Burial of Euclid, for instance, was apparently connected to an ancient Roman carnival.) Their songs and attitudes of mock reverence furthermore linked to other instances of cultural parodies, usually from the stage, as well as timely references to minstrelsy and music halls. Toasting math became a way of engaging with the popular culture of the day, especially theatrical productions.

The Trig Ceremonies at Vassar, from chapter 4, were the clearest instance of a theatrical math production. According to fictional stories inspired by actual events, they involved costumes, lighting, and sets, in addition to actors, directors, and scripts, all managed and supplied by the students in the sophomore class.[15] Some might say that the math content of these productions was perhaps less important than the collective action of coming together and putting on a show. Yet the reviews certainly did note the component, especially connecting it to the traditions of burning books at men's colleges. Making fun of/ with math became a topical touchstone for conversations among college students about what college meant.

Bringing these linkages back to the math classroom seems less common. Nicholas Murray Butler at Columbia, John Hudson Peck at Hamilton, and N. W. T. Root (from his partial time) at Yale all were prolific writers who participated in the Burial of Euclid or its iterations in their various campuses. Yet even when they admitted to participating in these "festivities," they did not expand on the broader implications for understanding math through theater.

Some of the math professors and teachers who saw these traditions—Jeremiah Day at Yale, Charles Davies at West Point and Trinity, William Smyth at Bowdoin, Oren Root at Hamilton, Achsah Ely at Vassar, and many others—wrote extensively about education without bringing their students' dramatic outlets back to the classroom. Even beyond the (sometimes negative) attitudes toward math expressed in these productions, what the students called "dramatics" was a student-centered activity—a matter for the student union rather than the classroom—and nineteenth-century theater often suffered from a bad reputation, as a corrupting influence and certainly a suspicious career choice.[16] Certainly these student traditions did not seem the right topic to bring back to academics.

Only recently have performance/theatrical exercises entered the math classroom. With support from the Alan Alda Center for Communicating Science, math teachers and researchers have begun to integrate techniques of improvisation.[17] Some, through the d.school (the Institute of Design at Stanford) or other advocates of design thinking, have argued for the importance of storytelling in design processes and talked about its implementation.[18] This book adds to these perspectives in pointing out that there is a long history of integrating American math and performance.

## Math Experiences and the Role of Gender

Given that students and alumni consistently said these performances were about their hatred of math, what other math attitudes have been reflected here? It seems clear that gleeful celebration and displays of academic achievement were factors in informal math events—as well as less desirable behaviors such as the destruction of school property, the excessive consumption of alcohol, and instances of exclusion. In other ways, in diaries, rebellions, speeches, and books, students noted their worries, their growing confidence, their enjoyment, their disdain, their thoughts about their teachers, and their sense of the value or uselessness of mathematics. Textbook authors, teachers, professors, presidents, and educational researchers meanwhile spun out their own tales about why math should be studied, why it should be useful for society, and even why it should be considered fun. This book has emphasized moments where attitudinal shifts come together, where even students' traditions of textbook destruction can be seen as rehearsals for teachers' and textbooks' rationalizing. In doing so, it has amassed a lot of historical data that seems to confirm that gender has played a major factor in shaping attitudes toward mathematics.

There is a long legacy of making assumptions about math studenthood and gender performance. The textbooks of Day and Davies and even the ones used in the early days of Vassar, Wells, and the Rutgers Female College all assumed that students would be male. N. W. T. Root provided a particularly striking example not only when he recommended shirtless mathematical calisthenics but also when he generated the slogan "Every Teacher *His* Own Drill-Master."[19] Similarly, the rules for oral communication at the blackboard assumed development into male officer-engineers, and the advice for teachers about cultivating a classroom environment addressed them with male pronouns. The language of arithmetic "drills" alongside military drills reified the martial environment, even before many educators advocated for women's participation in such spaces.

Throughout the nineteenth century, math's association with maleness became less evident as American classrooms did in fact become more diverse. As written tests replaced boardwork, they continued to assume obedience and discipline without explicit diatribes about "manly" behavior. Arithmetic drills developed from a certain pedagogical environment of showing off male bodies and mathematical minds—yet many people instead remembered their roots in Horace Mann's workshops that (revolutionarily for the time) catered to teachers of both sexes. Such activities certainly emphasized the role of competition in math, and so too did the speed tests—even sometimes retaining the reward of leaving class early for finishing. Still, few argued for math competition as a uniquely male attribute, even during the height of the American eugenics movement when certain understandings of Darwinian "sexual selection" could have supported such a claim.[20] Though the explicit studies of gender performance in math classrooms came later, they should perhaps give us pause even when looking at history.[21] Did certain aspects of math learning at that time (including but not limited to test design) assume male gender performance? At a time of explicit quotas regarding the presence of women and minorities in educational settings, were there aspects of math classrooms that acted as implicit barriers for their advancement?

Surprisingly, gender has fallen out of our discussions about math attitudes and experiences. Consider, for instance, a short history of quantifying math attitudes. In the 1950s, following the growth of educational research and written testing, education professor Wilbur Dutton proposed that preservice teachers pay attention to "measuring attitudes toward arithmetic," starting with themselves.[22] Some of the anecdotes he included were strikingly gendered. For instance, he quoted one education student at the University of California at Los Angeles: "Mother drilled and drilled me on arithmetic each morning while she combed my hair before going to school. I always went to school dreading arithmetic and any more mental punishment."[23] Though the gender of the speaker was not explicitly revealed, his anecdote (which he called "especially descriptive") seemed to suggest a female speaker or at least a female inquisitor (the mother). Still, the "objectivity" of Dutton's scale—its emphasis on quantitative measurement—was far more important for him than considerations of gender. When psychologist Lewis Aiken suggested additions in the 1960s and 1970s, he too emphasized the importance of separating Dutton's into multiple attitudinal scales: measuring what he called "enjoyment" and "value."[24] Aiken too did not address gender as an important factor.

Today's most cited tool, however, introduced the concern that math could be considered a "male domain." In the late 1970s, mathematician Elizabeth Fennema and psychologist Julia Sherman built on the previous attitude scales, except they expanded the tool to over a hundred questions in nine categories.[25] Called the "Fennema-Sherman Mathematics Attitudes Scales," their tool combined measurements (asserted to be related but separable) of the attitude toward success in mathematics (AS); mathematics as a male domain (MD); mother's attitudes (M); father's attitudes (F); teacher's attitudes (T); confidence in learning mathematics (C); mathematics anxiety (A); effectance motivation (i.e., feelings of competence/involvement; E); and beliefs about the usefulness of mathematics (U).[26] Fennema and Sherman wished to study the national trend that fewer girls than boys were electing math when given a choice. Their work proved foundational for studies of math anxiety as well as national debates about math curricula and instructional choices.[27] Though the impetus for these gender analyses has remained, now given the homely name of the "leaky pipeline," pieces of their tool have fallen out of favor.

There has been little discussion of the reasons modifications to the Fennema-Sherman Mathematics Attitudes Scales have dropped the gender component. When educational researcher Janet Melancon, educational psychologist Bruce Thompson, and math teacher Shirley Becnel proposed alterations in the 1990s, they claimed their alterations came from the intervening reliability studies and the need to make the test shorter for elementary school children.[28] Irish researchers Fiona Mulhern and Gordon Rae made similar arguments, as did mathematician Martha Tapia and educational researcher George Marsh II when they proposed alterations for studies of middle schoolers and high schoolers in the early 2000s.[29] Though such modifications have cut the test (and the time devoted to it) nearly in half, they have favored the retention of scales devoted to teacher's attitudes; confidence; anxiety; motivation; value; and enjoyment. The dropping of the male domain scale has not been explained, except when Tapia and Marsh claimed "gender did not have an effect on attitudes toward mathematics."[30]

In such scholarship and beyond, there has been a delicate balance between neglecting the roles of gender in math attitudes, on the one hand, and reinforcing stereotypes, on the other. Insisting too much that math could be seen as a "male domain" might introduce the worry for students who might otherwise feel included in math classes. As British sociologist Paul Ernest has said, educators do not wish to create a cycle whereby studying women's electing out of math makes it so.[31] That said, ignoring the role of gender in shaping educational experiences seems misguided, especially when considering such a powerful history. This book urges us to consider how the historical frame itself—telling stories of past successes, challenges, and general attitudes—can help introduce questions of the gendering of mathematical success while retaining a certain distance from today's classroom. It should be pointed out, for instance, that these stories of math communication have shown remarkable gender fluidity during a period in American history known for its reification of separate masculine/feminine expectations.

In short, this book reveals surprising complexity that goes beyond aspects of students and educators that can be quantitatively measured. The humor of students' informal activities, teachers' expectations for students' behavior, the dependence of classroom experiences on the communication technologies available, the incidence of rebellions big and small, the costuming of math learners, and the changes to classroom architecture: all these factors are not so straightforward, especially in light of psychometric studies of math anxieties and attitudes. As technical communicator Beth Flynn has argued, the quantitative measurement of feelings about quantification is perhaps the wrong way to approach the question.[32] In fact, some of the stories contained in this book can instead support the celebration of difference, ambiguity, and even ambivalence with regard to learning mathematics. What would math attitudes look like if math were assumed to be performative?

## Math Anxiety through Performance

It is clear that math classes have provoked discomfort for over two centuries. Yet the classroom experiences changed drastically through time as expectations of math communication went through shifts and reorientations. The image of hundreds of students writing on slates, their heads bent down, gave way to illustrations of boardwork: students speaking in front of blackboards. Oral communication replaced written—and then back again—as educational fashions changed. What do these changes have to do with math anxiety and tips for learning math today? We can get there by first thinking about some implications of considering math and performance together, paying attention to what has been shown and displayed.

Despite the new emphasis on public speaking in math classrooms, sometimes, the image of boardwork has emphasized silence instead. Winslow Homer's 1877 painting *The Blackboard* depicts a young woman in profile, facing a blackboard yet turned to her right. Her left hand holds her right arm, hugging her back. The blackboard contains multiple geometrical shapes: a right angle, a right triangle, and an equilateral triangle on the left; and parallel lines, a circle, a semicircle, and most of a square on the right. A line (perhaps on the blackboard, perhaps a long pointer) connects her right hand to the circle. Though the painting contains a hodgepodge of geometrical content, the woman is clearly not speaking, since her mouth is closed.[33] The painting, though in private ownership for nearly a hundred years, has been on display often in the U.S. National Gallery of Art since 1990, and some of its uses have encouraged children to see themselves in the painting.[34] It has suggested that the common school experience is not so much speaking in front of a blackboard as staying silent in front of one.

Though math classrooms came to be depicted as sites of writing once again with the spread of preprinted testing, common images did not show public writing as much as collective attempts at secrecy. The image of the blackboard—whether as a site for speaking or a site for silence—came to be replaced with depictions of rows of students taking high-stakes, timed tests. Boardwork had included writing, too, of course; the young woman in Homer's painting occupies the same canvas as circles and squares.[35] But the status of the writing changed. In competitive testing environments, attempts at answers were not supposed to be shown: students were directed to keep their tests hidden from their classmates. Though depicting collective writing, images of testing therefore have emphasized what is not revealed, what is secret, instead.

In the case of learning math, performance encourages us to pay attention to both what is there and what is not. Performance—particularly in the sense of theatrical performance—gives the semblance of real action to (usually) prewritten lines, providing a frame through which to see other people. Though it seems to be about speech and display, theater also encourages us to notice instances of silence and absence.[36] Along these lines, math communication has been about not only speaking and writing but also barriers to these actions.

Such considerations reinforce certain aspects of math anxiety. Math anxiety, medically, is a condition that prevents communicative performance. A pounding heartbeat, difficulty breathing, tension, and pain get in the way of "doing" math—usually, of speaking and writing about mathematics.[37] Math anxiety results in moments of silence and hiding. Its presence follows from the performative connections to math education, just as the actions of reading, speaking, writing, rehearsing, and acting do. Not only has stage fright been widespread, but so too has performances' handling of absence and omission. A performative appreciation of math would mix instances of breakdowns as well as successful communication.

Through its organization, *Performing Math* has indicated the interplay between math anxiety and math communication. Worries about bodily display have joined conversations about textbooks, reading, and quantitative literacy. Expectations for public speaking at the blackboard have inspired rebellions, textbook burials, and doodles. Unease over women's math classes has led to theatrical productions and news editorials. Worries about written testing have fed the further development of such tests and other quantitative measurements of students. From the initial faculty response regarding the Conic Sections Rebellion to the development of MARS, it has seemed normal to respond to discomfort over math with even more mathematics. Math anxiety and math communication have reinforced each other.

This book therefore does not offer a comprehensive list of suggestions for dealing with math anxiety. Such tips can be borrowed from stories of stage fright, from the work other people have done in reflecting on and consolidating their experiences. Instead, the book offers a "reorientation"—in the words of McCleary and McKinney.[38] In suggesting comparisons with stage fright, it makes math anxiety seem more unpredictable, more pervasive, and more serious than previously imagined. It is not just a matter of situational anxiety that affects the performance of calculation; it can occur for a large part of the American population in many different situations. Moreover, the general comparisons with performance show how math communication and math anxiety should be considered together.

Still, such an argument (appreciating the continued connections between math and performance) does not suggest an entirely static view. Performance, after all, is often a living medium that develops over time, depending on people gathering together. There is tremendous variety. Math experiences might sometimes seem like a Shakespeare play, at others like a Broadway show. They might have components of a circus, a medieval miracle play, or a rock concert. Classrooms might feel like sites of Gilbert and Sullivan's parody or Samuel Beckett's tragedy. The expectations for our behavior might feel as set as the lines of ancient drama or as fresh as Second City improv. We can begin to recognize that math is not static, that it depends on collaborative human actions.[39]

Moreover, the performative components of math education allow us to see connections to broader questions in education studies. How might we better frame studies of education around student experience, especially their affective experience? In other words, what might mathematical studies look like if we took seriously the instances of math hatred or at least math anxiety? Second, how does the "justification problem" for mathematics change contextually, and what might that mean, for instance, for local arguments that mathematics should be studied because it provides a "global" language? Lastly, most broadly, what are the connections between performing mathematics and the creep of performance assessment in business, industry, education, and beyond? What historically connects these realms? In sum, *Performing Math* does not just prove the argument that math education has performative components; such connections also make possible the introduction and investigation of much bigger issues.

Certain educational movements since 2000 have been emphasizing these performative connections. The Alan Alda Center for Communicating Science has developed the mission of "empower[ing] scientists and health professionals to communicate complex topics in clear, vivid, and engaging ways," promising to combine perspectives on STEM and health with "journalism, communications, public policy and theater."[40] Their particular mix of storytelling and improv techniques has been a notable addition to the STEM communicators' toolkit. Because their examples sometimes seem to emphasize women who need help and men who provide models of success, certain aspects of their framework seem to reassert some of the assumptions that Fennema and Sherman studied forty years ago.[41] Still, their constant advocacy for theater as a portion of STEM communication has been valuable and exciting.

Certain advocates of design thinking, too, have made similar arguments recently. Their suggested processes have aimed to foster innovation by promoting empathy, understanding potential users, engaging in wild brainstorming, and supporting rapid prototyping. Many of their techniques implicitly have assumed a certain performativity, and some explicitly have suggested the techniques of acting (especially improv) to develop engaging storytelling.[42] Though their early years saw the emergence of white male engineers as their primary spokespeople, design thinking workshops have attempted to include more women. Combined with perspectives from the Center for Communicating Science, the movement has argued for the importance of supporting views from the arts (especially theater) so that various organizations can better encourage creativity and new ideas.

In asserting the importance of considering performance and math communication together, this book offers a reorientation to these educational movements as well. Though math is contained in certain examples from Alan Alda's team, the Center for Communicating Science emphasizes helping health professionals and scientists. Similarly, though math experiences are supposed to benefit from design thinking, much of their materials assume frameworks of engineering. Instead of emphasizing the neglect of math in recent movements, the historical cases in this book urge consideration of the interdisciplinary/ multidisciplinary relationships that make up STEAM.[43] American math, after all, already has been performative.

In sum, learning math has been like putting on a show. From the perspectives of math communication, it has started with reading aloud, it has been practiced in prohibited ways, and it has been literally theatrical. From the perspectives of math anxiety, worries have developed from classroom tech and have become about written testing. Moreover, the overall framework of learning math, represented in the five chapters of this book, can be considered as analogous to the process of developing a theatrical production: from read-through to early rehearsals, tech/dress week, opening night, and house management. (Perhaps look back at the five chapters of this book through the lens of theatrical process.) Appreciating performative connections can leave us with frameworks for old classroom activities that might be new again, arguments for pedagogical innovations, broad-based reorientations, and even perspectives on STEAM movements. Overall, *Performing Math* suggests that learning math has depended on human actors and collective actions, that it has changed over time in changing spaces, and that it has been exciting, terrifying, emotional, and affective—in short, that learning math has been performative and even dramatic for millions of Americans.

# Acknowledgments

Since *Performing Math* is my first book, it has received a lot of support over many years. I have so many people to thank; I will not be able to name them all. I will start by acknowledging the support of my extended family: the Fisses, Rodriguezes, Mirandas, Kassons, Franc family, and many others. Their enthusiasm has meant so much. I am especially grateful for productive conversations with my parents-in-law, Joy Kasson and John Kasson, and for some close reading from my parents, Willie Fiss and Mike Fiss. This book is dedicated to my wife, the inimitable scholar Laura Kasson Fiss, and our two sons, Sebastian and Toby. All the people mentioned previously have acted as inspiring teachers of our sons, but I especially need to mention their actual teachers at Little Huskies Child Development Center, who have provided amazing childcare and some of our best experiences as a family.

I have also had the good fortune to be part of many exciting educational communities. Thanks to the Vassar Department of Mathematics and Statistics, especially John McCleary, for providing such a welcoming introduction to college-level mathematics. I continued to be engaged through graduate work in Indiana University's History and Philosophy of Science and Medicine, especially under Sandy Gliboff, Jim Capshew, and Jutta Schickore. Thanks to my student cohorts at both Vassar and Indiana for companionship and fun. When I completed my PhD, I had the good fortune to go back to Vassar, where Janet Gray helped me participate in the programs in science, technology, and society and women's studies. Much of the archival work contained in this book comes from my second stay at Vassar because my position provided generous funds for research travel through the Andrew W. Mellon Foundation. Next, I had a brief, though transformative, stay at Davidson College, where Van Hillard provided excellent mentorship. Now I am lucky to find an engaging community in Michigan Technological University's Department of Humanities, especially their undergraduate program in scientific and technical communication and graduate program in rhetoric, theory, and culture. Thanks especially to Steve Walton, Ann Brady, and Vicky Bergvall for welcoming me to this part of the world.

Even beyond individual institutions, research communities have provided invaluable feedback throughout the making of this book. Karen Parshall and the Forum for the History of the Mathematical Sciences provided early and enthusiastic discussion, as did Kidwell, Roberts, Ackerberg-Hastings, and the History and Pedagogy of Mathematics—Americas Section. More recently, I have been very lucky to present this research at symposiums in the United Kingdom and the United States. So many thanks to Amy Chambers for her vision and voice, for arranging the *Mathematikado* event at the British Science Festival, and for inviting me and Laura to present our work. Thank you to Morgan Ames, Massimo Mazzotti, and the other participants of the Algorithmic Thinking symposium at the University of California at Berkeley for inspiring conversation and productive feedback, as to Michael Barany and the other participants in the British Academy's Universals' Locales workshop at the University of Edinburgh.

Librarians, archivists, and editors have been essential to this project, as they are to any research. For their insights and archivist skill, thank you to Kat Stefko at Bowdoin College, Elaine Ardia at Bates College, Susan Lintelmann and Alicia Mauldin-Ware at the U.S. Military Academy, and Ron Patkus and Dean Rogers at Vassar College. For their friendship and advice, thanks too to Gretchen Lieb at Vassar, Martha Harsanyi at Indiana University, Matthew Vest when at Davidson, and Erin Matas at Michigan Tech. As for editors, this book benefited from some early feedback from Bill Germano, sustained advice from Jeff Dean, and now the amazing attention of Lisa Banning. Lisa Banning, especially, has made the publishing process impressively quick, intensive, and rewarding.

As I list my thanks, I realize Laura needs to be mentioned again. Not only has she provided amazing advice and friendship throughout our travels through many states and years, but she has also been part of many of the educational and research communities listed previously. Her research has shaped so much of what's good here, from *The Mathematikado* and beyond, and I am excited to continue presenting and publishing with her.

In sum, so many people have shaped this book in so many good ways. Any errors are mine.

# Notes

## Preface

1 Heather Mendick, "A Beautiful Myth? The Gendering of Being/Doing 'Good at Maths,'" *Gender and Education* 17, no. 2 (2005): 203–219; and Heather Mendick, "Mathematical Stories: Why Do More Boys Than Girls Choose to Study Mathematics at AS-Level in England?," *British Journal of Sociology of Education* 26, no. 2 (2005): 235–251.

2 In our textbook, Edward D. Gaughan, *Introduction to Analysis*, 5th ed. (Pacific Grove, Calif.: Brooks/Cole, 1998), 34.

3 John McCleary, *Exercises in (Mathematical) Style: Stories of Binomial Coefficients* (Washington, D.C.: MAA, 2017).

4 The common example can, for instance, be found in David P. Rosin, *Dynamics of Complex Autonomous Boolean Networks* (Dordrecht, Neth.: Springer Theses, 2015), 58.

5 Sara Hottinger, *Inventing the Mathematician: Gender, Race, and Our Cultural Understandings of Mathematics* (Albany, N.Y.: SUNY Press, 2016), 1–2.

6 Mendick, "A Beautiful Myth?"; Mendick, "Mathematical Stories"; and Valerie Walkerdine, *Counting Girls Out* (London: Routledge, 1998).

7 Jae Hoon Lim, "Double Jeopardy: The Compounding Effects of Class and Race in School Mathematics," *Equality and Excellence in Education* 41, no. 1 (2008): 81–97.

8 Alan Alda, *If I Understood You, Would I Have This Look on My Face? My Adventures in the Art and Science of Relating and Communicating* (New York: Random House, 2016).

9 Susan Greenfield, "A New Kind of Literacy," *Guardian*, 9 April 2003.

10 Andrew Hacker, *The Math Myth and Other STEM Delusions* (New York: New Press, 2016), 11.

11 Some records of Vassar's Trig Ceremonies appear in "V.C. Trig Ceremonies," subject files folder 25.35, Archives and Special Collections Library, Vassar College.

12 For instance, Stephen Harris et al., "17: Science by the Seaside," 13 September 2017, in *Anthill*, produced by The Conversation UK, podcast, MP3 audio, 35:40, accessed 8 September 2018, http://theconversation.com/anthill-17-science-by-the-seaside -84008.

13 See Hottinger, *Inventing the Mathematician*.
14 See "The Mathematikado: The Mystical 'Trig Ceremonies' of Sweet Girl Graduates," *Cornell Daily Sun*, 7 April 1886; and "The Mathematikado: 'Trig Ceremonies' at Vassar," *Harvard Crimson*, 25 March 1886.

## Introduction

1 See Andrew Fiss, "Mathematics and Mourning: Textbook Burial and Student Culture before and after the Civil War, 1853–1880," *History of Education Quarterly* 57, no. 2 (2017): 221–246. For earlier, incomplete references to the practice, see Alexandra Robbins, *Secrets of the Tomb: Skull and Bones, the Ivy League, and the Hidden Paths of Power* (Boston: Little, Brown, 2002), 31–33; Laurie A. Wilkie, *The Lost Boys of Zeta Psi: A Historical Archaeology of Masculinity at a University Fraternity* (Berkeley: University of California Press, 2010), 60, 84, 98, 127; Robert V. Bruce, *The Launching of Modern American Science, 1846–1876* (New York: Knopf, 1987), 85–86; Stanley Guralnick, *Science and the Ante-bellum American College* (Philadelphia: American Philosophical Society, 1975), 59; Ryan K. Anderson, "'The Law of College Custom Is [as] Inexorable as the Laws of Chemistry and Physics': The Transition to a Modern Purdue University, 1900–1924," *Indiana Magazine of History* 99, no. 2 (2003): 97–128; and Helen Lefkowitz Horowitz, *Campus Life: Undergraduate Cultures from the End of the Eighteenth Century to the Present* (Chicago: University of Chicago Press, 1988), 34.
2 Judith Butler, *Gender Trouble: Feminism and the Subversion of Identity* (New York: Routledge, 2008 [1990]), 34.
3 Delphine Gardey, "Writing the History of the Relations between Medicine, Gender and the Body in the Twentieth Century: A Way Forward?," *Clio* 37 (2013), https://journals.openedition.org/cliowgh/404. She (or her translator) expands SSK as "social studies of knowledge," which is not standard.
4 Thomas S. Kuhn, *The Structure of Scientific Revolutions* (Chicago: University of Chicago Press, 1970 [1962]), 45–46. My particular reading also owes a lot to Michael Polanyi, *Personal Knowledge: Towards a Post-critical Philosophy* (Chicago: University of Chicago Press, 1958), 60.
5 See Jan Golinski, *Making Natural Knowledge: Constructivism and the History of Science* (Chicago: University of Chicago Press, 2005 [1998]), 17–46.
6 John McCleary and Audrey McKinney, "What Mathematics Isn't," *Mathematical Intelligencer* 8, no. 3 (September 1986): 51–53.
7 Michael J. Barany and Donald MacKenzie, "Chalk: Materials and Concepts in Mathematics Research," in *Representation in Scientific Practice Revisited*, ed. Catelijne Coopmans, Janet Vertesi, Michael E. Lynch, and Steve Woolgar (Cambridge, Mass.: MIT Press, 2014), 107–130.
8 See definition 1 in *OED Online*, s.v. "performance, n.," last modified March 2019, https://www.oed.com/view/Entry/140783.
9 See Karen Hunger Parshall and David E. Rowe, *The Emergence of the American Mathematical Research Community, 1876–1900: J. J. Sylvester, Felix Klein, and E. H. Moore* (Providence, R.I.: American Mathematical Society, 1994); Theodore M. Porter, *Trust in Numbers: The Pursuit of Objectivity in Science and Public Life* (Princeton: Princeton University Press, 1995); Kim Tolley, *The Science Education of American Girls: A Historical Perspective* (London: RoutledgeFalmer, 2003); Wilkie, *Lost Boys*; and Michael David Cohen, *Reconstructing the Campus: Higher*

*Education and the American Civil War* (Charlottesville: University of Virginia Press, 2012).

10 The example appears in Mike Markel and Stuart A. Selber, *Technical Communication*, 12th ed. (Boston: Bedford / St. Martin's, 2018), 457–464.

11 Such an approach can be seen in the growth of "engineering communication," such as Charles W. Knisely and Karin I. Knisely, *Engineering Communication* (Boston: Cengage, 2014).

12 Carolyn Miller, "A Humanistic Rationale for Technical Writing," *College English* 40, no. 6 (1979): 610–617.

13 Katherine T. Durak, "Gender, Technology, and the History of Technical Communication," *Technical Communication Quarterly* 6, no. 3 (1997): 249–260.

14 Lara Zarum, "Math Class Is Tough, but Fixing Barbie Is Harder," *Village Voice*, 26 April 2018. Also see Jonathan H. Liu, "Echoes of Barbie's 'Math Class Is Tough!,'" *Wired*, 16 December 2011; and John Boone, "The 14 Most Controversial Barbies Ever," *Entertainment Tonight Online*, 24 November 2014.

15 William Harms, "When People Worry about Math, the Brain Feels the Pain," UChicago News, 31 October 2012. The university news story points back to the following research article: Ian M. Lyons and Sian L. Beilock, "When Math Hurts: Math Anxiety Predicts Pain Network Activation in Anticipation of Doing Math," *PLoS One* 7, no. 10 (2012): e48076.

16 Cathy O'Neil, *Weapons of Math Destruction: How Big Data Increases Inequality and Threatens Democracy* (Lake Arbor, Md.: Crown Books, 2016).

17 Elizabeth A. Flynn, "Feminism and Scientism," *College Composition and Communication* 46, no. 3 (October 1995): 353–368.

18 Davida Charney, "Empiricism Is Not a Four-Letter Word," *College Composition and Communication* 47, no. 4 (1996): 567–593.

19 Steven B. Katz, "The Ethics of Expediency," *College English* 54, no. 3 (1992): 255–275.

20 C. P. Snow, *The Two Cultures and the Scientific Revolution* (Cambridge: Cambridge University Press, 1961), 10–11.

21 From C. P. Snow, *The Two Cultures: A Second Look* (Cambridge: Cambridge University Press, 1963).

22 *Good Will Hunting*, directed by Gus Van Sant (Miramax Films, 1997).

23 *A Beautiful Mind*, directed by Ron Howard (Imagine Entertainment, 2001); and *Proof*, directed by John Madden (Lionsgate, 2005).

24 *Hidden Figures*, directed by Theodore Melfi (Twentieth Century Fox, 2016).

25 George Sarton, *The Study of the History of Mathematics and the Study of the History of Science* (New York: Dover, 1936), 8.

26 [B. H. Hall], *A Collection of College Words and Customs* (Cambridge, Mass.: John Bartlett, 1851), 26–32.

27 Andrew Hacker, *The Math Myth and Other STEM Delusions* (New York: New Press, 2016).

28 Milly S. Barranger, *Theatre: A Way of Seeing*, 4th ed. (Belmont, Calif.: Wadsworth, 1995), 2–24.

29 See *OED Online*, s.v. "math, n.3," last modified December 2019, https://www.oed.com/view/Entry/114962.

30 "Descriptive on the Brain," in Paul Dahlgren's diary, *Private Journal. No. 5*, folder 2238, Nineteenth Century Manuscripts Collection, Special Collections and Archives Division, United States Military Academy Library.

31 See Paul Cobb, Erna Yackel, and Kay McClain, *Symbolizing and Communicating in Mathematics Classrooms: Perspectives of Discourse, Tools, and Instructional Design* (Mahwah, N.J.: Lawrence Erlbaum, 2000); David Pimm, *Speaking Mathematically: Communication in the Mathematics Classroom* (London: Routledge, 1987); Sheila Tobias, *Overcoming Math Anxiety* (New York: Norton, 1978); and Sheila Tobias, *Succeed with Math: Every Student's Guide to Conquering Math Anxiety* (New York: College Entrance Examination Board, 1987).

32 McCleary and McKinney, "What Mathematics Isn't," 51–53.

33 Peggy Aldrich Kidwell, Amy Ackerberg-Hastings, and David Lindsay Roberts, *Tools of American Mathematics Teaching, 1800–2000* (Washington, D.C.: Smithsonian Institution / Baltimore, Md.: Johns Hopkins University Press, 2008).

34 Kidwell, Ackerberg-Hastings, and Roberts, xi–xiv.

35 See Christopher G. White, *Other Worlds: Spirituality and the Search for Invisible Dimensions* (Cambridge, Mass.: Harvard University Press, 2018); Parshall and Rowe, *Mathematical Research Community*; Porter, *Trust in Numbers*; Cohen, *Reconstructing the Campus*; and Wilkie, *Lost Boys*.

## Chapter 1   How Math Communication Has Started with Reading Aloud

1 See the "Conic Rebellion 1830 Statements and Drafts of Statements," folder YRG 41-C 1–8 in RU 83, box 1, Yale University Archives, Manuscripts & Archives, Sterling Memorial Library, Yale University.

2 "Conic Rebellion 1830 Petitions and Dismissals," folder YRG 41-C 1–9 in RU 83, box 1, Yale University Archives.

3 Burton J. Bledstein, *The Culture of Professionalism: The Middle Class and the Development of Higher Education in America* (New York: W. W. Norton, 1976), 231.

4 Henry D. Sheldon, *Student Life and Customs* (New York: Appleton, 1901), 110.

5 Brooks Mather Kelley, *Yale: A History* (New Haven, Conn.: Yale University Press, 1974), 168–169; Clarence Deming, "Yale Wars of the Conic Sections," *Independent* 56 (1904): 667–669; and William Lathrop Kingsley, *Yale College: A Sketch of Its History, with Notices of Its Several Departments, Instructors, and Benefactors, Together with Some Account of Student Life and Amusements* (New York: H. Holt, 1879), 1:137–138.

6 Peggy Aldrich Kidwell, Amy Ackerberg-Hastings, and David Lindsay Roberts, *Tools of American Mathematics Teaching, 1800–2000* (Washington, D.C.: Smithsonian Institution / Baltimore, Md.: Johns Hopkins University Press, 2008), 25–26.

7 E. Jennifer Monaghan, *Learning to Read and Write in Colonial America* (Amherst: University of Massachusetts Press, 2005); Patricia Cline Cohen, *A Calculating People: The Spread of Numeracy in Early America* (Chicago: University of Chicago Press, 1982); and N. Ellerton and M. A. K. Clements, *Rewriting the History of School Mathematics in North America, 1607–1861: The Central Role of Cyphering Books* (Dordrecht, Neth.: Springer, 2012).

8 A National Educational Association report from 1894 criticized arithmetic classes precisely for this legacy. See Andrew Fiss, "Problems of Abstraction: Defining an American Standard for Mathematics Classes at the Turn of the Twentieth Century," *Science & Education* 21, no. 8 (2012): 1185–1197.

9 Jeremiah Day, *An Introduction to Algebra, Being the First Part of a Course of Mathematics, Adapted to the Method of Instruction in the American Colleges* (New Haven,

Conn.: Howe & Deforest, 1814), 3. A copy consulted is in the Lilly Library, Indiana University.

10 Day, 3–6.

11 Day, 3.

12 Day, 3–4.

13 The earliest number in a continuous breakdown is an estimate for 1869–1870: 1.3 percent of the eighteen- to twenty-four-year-old population attended college. See Thomas D. Snyder, ed., *120 Years of American Education: A Statistical Portrait* (Washington, D.C.: National Center for Education Statistics, 1993), 76. This was perhaps the wrong age range for the time.

14 See, for instance, Kim Tolley, *The Science Education of American Girls: A Historical Perspective* (London: RoutledgeFalmer, 2003), 6. The same students could very well attend institutions called *colleges, schools,* or *academies.* Ellerton and Clements, *Rewriting Mathematics,* 2–3.

15 Craig Steven Wilder, *Ebony and Ivy: Race, Slavery, and the Troubled History of America's Universities* (New York: Bloomsbury, 2013), 149–180; and Jon Reyhner and Jeanne Eder, *American Indian Education: A History* (Norman: University of Oklahoma Press, 2004), 29–33.

16 See Andrew Fiss, "Cultivating Parabolas in the Parlor Garden: Reconciling Mathematics Education and Feminine Ideals in Nineteenth-Century America," *Science & Education* 23, no. 1 (2014): 241–250.

17 Crystal Broch Colombini and Sue Hum, "Integrating Quantitative Literacy into Technical Writing Instruction," *Technical Communication Quarterly* 26, no. 4 (2017): 379–394.

18 Colombini and Hum, 380.

19 Nathan Grawe and Carol Rutz, "Integration with Writing Programs: A Strategy for Quantitative Reasoning Program Development," *Numeracy* 2, no. 2 (2009): 1–18.

20 The most influential, recent statement of the mental discipline argument is in Piers Bursill-Hall, *Why Do We Study Geometry? Answers through the Ages* (Cambridge: Centre for Mathematical Sciences at the University of Cambridge, 2002), 1–31.

21 Adrian Smith, *Making Mathematics Count: The Report of Professor Adrian Smith's Inquiry into Post-14 Mathematics Education* (London: HMSO, 2004).

22 See Paul Chambers and Robert Timlin, *Teaching Mathematics in the Secondary School* (Los Angeles: Sage, 2013). For a slightly older statement, see H. Freudenthal, *Mathematics as an Educational Task* (Dordrecht, Neth.: D. Reidel, 1973).

23 Day, *Introduction to Algebra.*

24 John Locke, *Posthumous Works of Mr. John Locke* (London: W. B. for A. and J. Churchill at the Black Swan in Pater-Noster-Row, 1706).

25 Laurie Wilkie, *The Lost Boys of Zeta Psi: A Historical Archaeology of Masculinity at a University Fraternity* (Berkeley: University of California Press, 2010), 7–8.

26 George Washington Cullum, *Biographical Register of the Officers and Graduates of the U.S. Military Academy at West Point, N.Y. from Its Establishment March 16, 1802 to the Army Re-organization of 1866–67* (New York: D. Van Nostrand, 1868).

27 Wilkie, *Lost Boys,* 1–28.

28 Theodore Dwight Woolsey, "Article V.—President Woolsey's Address at the Funeral of President Day, Commemorative of His Life and Service," *New Englander and Yale Review* 26 (1867): 692–724; Ebenezer Baldwin, *Annals of Yale College, from Its Foundation, to the Year 1831* (New Haven, Conn.: B. & W. Noyes, 1838), 155–169; and Rossiter Johnson and John Howard Brown, eds., *The Twentieth*

*Century Biographical Dictionary of Notable Americans* (Boston: The Biographical Society, 1904), s.v. "Day, Jeremiah, educator."

29  David Lindberg, *The Beginnings of Western Science: The European Scientific Tradition in Philosophical, Religious, and Institutional Context, 600 B.C. to A.D. 1450* (Chicago: University of Chicago Press, 1992), 183–214. The specific construction of the liberal arts had its roots in the Roman encyclopedia *Nine Books of Disciplines*. See Elizabeth Rawson, *Intellectual Life in the Late Roman Republic* (Baltimore, Md.: Johns Hopkins University Press, 1985), 158–164.

30  Frederick Rudolph, *The American College and University: A History* (Athens: University of Georgia Press, 1991), 44–67; Paul Westmeyer, *A History of American Higher Education* (Springfield, Ill.: Charles C. Thomas, 1985), 23–60; and Paul Venable Turner, *Campus: An American Planning Tradition* (New York: Architectural History Foundation, 1987), 16–51.

31  Woolsey, "Address," 697–699.

32  Day, *Introduction to Algebra*, 3.

33  Day, 4–7.

34  Day, 5.

35  Day, 5.

36  Amy Ackerberg-Hastings contributed the textbook argument to the coauthored Kidwell, Ackerberg-Hastings, and Roberts, *American Mathematics Teaching*, 8–9. See also Amy Ackerberg-Hastings, "Mathematics Is a Gentleman's Art: Analysis and Synthesis in American College Geometry Teaching, 1790–1840" (PhD diss., Iowa State University, 2000).

37  Day, *Introduction to Algebra*; and Jeremiah Day, *A Treatise of Plane Trigonometry to Which Is Prefixed a Summary View of the Nature and Use of Logarithms, Being the Second Part of a Course of Mathematics, Adapted to the Method of Instruction in the American Colleges* (New Haven, Conn.: Howe & Deforest, 1815).

38  *A Practical Application of the Principles of Geometry to the Mensuration of Superficies and Solids, Being the Third Part of a Course of Mathematics, Adapted to the Method of Instruction in the American Colleges* (New Haven, Conn.: Oliver Steele, 1814).

39  *The Mathematical Principles of Navigation and Surveying, Being the Fourth Part of a Course of Mathematics, Adapted to the Method of Instruction in the American Colleges* (New Haven, Conn.: Steele & Gray, 1817).

40  Florian Cajori, *The Teaching and History of Mathematics in the United States* (Washington, D.C.: Government Printing Office, 1890), 63–64; Helena Pycior, "British Synthetic vs. French Analytic Styles of Algebra in the Early American Republic," in *The History of Modern Mathematics*, ed. David E. Rowe and John McCleary (Boston: Academic Press, 1989), 1:125–154; and Karen Hunger Parshall and David E. Rowe, *The Emergence of the American Mathematical Research Community, 1876–1900: J. J. Sylvester, Felix Klein, and E. H. Moore* (Providence, R.I.: American Mathematical Society, 1994), 15.

41  Harvey Green, *Fit for America: Health, Fitness, Sport, and American Society* (New York: Pantheon, 1986); and Wilkie, *Lost Boys*. For a similar case for women, see Helen Lefkowitz Horowitz, *Alma Mater: Design and Experience in the Women's Colleges from Their Nineteenth-Century Beginnings to the 1930s* (Amherst: University of Massachusetts Press, 1993).

42  Fiss, "Cultivating Parabolas," 241–250.

43 "Nathaniel William Taylor Root," in *Obituary Record of Graduates of Yale College, Deceased during the Academical Year Ending in June, 1873* (New Haven, Conn.: Yale University Press, 1873), 105; and Kentwood D. Wells, "Contraband Christmas," *Magic Lantern Gazette* 18, no. 4 (Winter 2006): 3–7.

44 N. W. Taylor Root, *School Amusements; or, How to Make the School Interesting* (New York: A. S. Barnes, 1857). The secret publication is [N. W. T. Root and J. K. Lombard], *Songs of Yale* (New Haven, Conn.: E. Richardson, 1853).

45 Root, *School Amusements*, 217–218.

46 Root, 217–221.

47 Root, 221–222.

48 Root, 221–222.

49 Root, 220.

50 Root, 11–148.

51 Charles Davies, *Practical Mathematics, with Drawing and Mensuration, Applied to the Mechanic Arts* (New York: A. S. Barnes, 1852). Charles Davies's influence on American school movements is further chronicled in Kidwell, Ackerberg-Hastings, and Roberts, *American Mathematics Teaching*, 4, 10–12, 15–20, 173, 185.

52 Root, *School Amusements*, 19–148.

53 Root, 96.

54 Root, 98.

55 Root, xi–xv.

56 See, for one reference among many, Toby Lester, *Da Vinci's Ghost: Genius, Obsession, and How Leonardo Created the World in His Own Image* (New York: Free Press, 2012).

57 For analyses along these lines, see Green, *Fit for America*; and Wilkie, *Lost Boys*.

58 For a recent book about "teaching math to boys," see Michael Reichert and Richard Hawley, *Reaching Boys, Teaching Boys: Strategies That Work—and Why* (San Francisco: Jossey-Bass, 2010).

59 Fiss, "Cultivating Parabolas."

## Chapter 2    How Math Communication Has Been Practiced in Prohibited Ways

1 For examples of such analysis, see Alexandra Robbins, *Secrets of the Tomb: Skull and Bones, the Ivy League, and the Hidden Paths of Power* (Boston: Little, Brown, 2002), 31–33; Robert V. Bruce, *The Launching of Modern American Science, 1846–1876* (New York: Knopf, 1987), 85–86; Stanley Guralnick, *Science and the Ante-bellum American College* (Philadelphia: American Philosophical Society, 1975), 59; and Ryan K. Anderson, "'The Law of College Custom Is [as] Inexorable as the Laws of Chemistry and Physics': The Transition to a Modern Purdue University, 1900–1924," *Indiana Magazine of History* 99, no. 2 (2003): 97–128.

2 Helen Lefkowitz Horowitz, *Campus Life: Undergraduate Cultures from the End of the Eighteenth Century to the Present* (Chicago: University of Chicago Press, 1987), 3–22.

3 Horowitz.

4 See Andrew Fiss, "Mathematics and Mourning: Textbook Burial and Student Culture before and after the Civil War, 1853–1880," *History of Education Quarterly* 57, no. 2 (2017): 221–246, https://doi.org/10.1017/heq.2017.3.

5 "Mathematicae Exsequiae, a Classe Juniore, Collegii Bowdoinensis, VIII. Kal. Augusti. A.D. 1854," in Programs for funerals of mathematics textbooks, Anna

Lytica + calculus, 1851–1879, Bowdoin Memorabilia and Realia folder 10.2, George J. Mitchell Department of Special Collections and Archives, Bowdoin College.

6  Nancy L. Hoft, *International Technical Communication* (New York: Wiley, 1995); and Elizabeth Tebeaux and Linda Driskill, "Cultural and the Shape of Rhetoric: Protocols of International Document Design," in *Exploring the Rhetoric of International Professional Communication* (Amityville, N.Y.: Baywood, 1999), 211–251.

7  James Lloyd Winstead, *When Colleges Sang: The Story of Singing in American College Life* (Birmingham: University of Alabama Press, 2013), 71.

8  Fiss, "Mathematics and Mourning."

9  Winstead, *When Colleges Sang.*

10  Michael S. Hevel, "A Historiography of College Students 30 Years after Helen Horowitz's *Campus Life*," in *Higher Education: Handbook of Theory and Research*, ed. Michael B. Paulsen (Dordrecht, Neth.: Springer International, 2017), 419–484; and Michael S. Hevel and Heidi A. Jaeckle, "Trends in the Historiography of American College Student Life: Populations, Organizations, and Behaviors," in Paulsen, *Higher Education*, 11–36.

11  [Ezekiel Porter Belden], *Sketches of Yale College, with Numerous Anecdotes, and Embellished with More than Thirty Engravings* (New York: Saxton & Miles, 1843), 166–170.

12  [B. H. Hall], *A Collection of College Words and Customs* (Cambridge, Mass.: John Bartlett, 1851), 27–33.

13  [N. W. T. Root and J. K. Lombard], *Songs of Yale* (New Haven, Conn.: E. Richardson, 1853).

14  [Root and Lombard], 28–30.

15  Winstead, *When Colleges Sang*, 82–98.

16  [Root and Lombard], *Songs of Yale*, 30.

17  [Root and Lombard], 28–29.

18  [Root and Lombard], 28.

19  [Root and Lombard], 3.

20  "Nathaniel William Taylor Root," in *Obituary Record of Graduates of Yale College, Deceased during the Academical Year Ending in June, 1873* (New Haven, Conn.: Yale University Press, 1873), 105.

21  "James Kittredge Lombard," in *Obituary Record of Graduates of Yale University, Deceased during the Academic Year Ending in 1890* (New Haven, Conn.: Yale University Press, 1890), 583–584.

22  Margaret A. Nash, Danielle C. Mireles, and Amanda Scott-Williams, "'Mattie Matix' and Prodigal Princes: A Brief History of Drag on College Campuses from the Nineteenth Century to the 1940s," in *Rethinking Campus Life: New Perspectives on the History of College Students in the United States*, ed. Christine A. Ogren and Marc A. VanOverbeke (New York: Palgrave Macmillan, 2018), 61–90.

23  Laurie A. Wilkie, *The Lost Boys of Zeta Psi: A Historical Archaeology of Masculinity at a University Fraternity* (Berkeley: University of California Press, 2010), 60, 84, 98, 127.

24  Marjorie Garber, *Vested Interests: Cross-Dressing and Cultural Anxiety* (New York: Routledge, 1997).

25  Frederick S. Hills, ed., "John Hudson Peck," in *New York State Men: Biographic Studies and Character Portraits* (Albany, N.Y.: Argus, 1910), 102–103.

26  John Hudson Peck, "Technical Education," *Journal of the Western Society of Engineers* 2 (1897): 231.

27  John Hudson Peck, "Reunion of the Class of 1859, June 23, 1909," in *Hamilton College Half-Century Annalist's Letter* (Clinton, N.Y.: Hamilton College, 1909).

28  Peck.

29  Peck.

30  Peck.

31  Peck.

32  Peck.

33  Peck.

34  John F. Kasson, *Civilizing the Machine: Technology and Republican Values in America, 1776–1900* (New York: Grossman, 1976), 3–51; and Ruth Schwartz Cowan, *A Social History of American Technology* (Oxford: Oxford University Press, 1997), 67–220.

35  Elihu Root, "The Centenary of Hamilton College: Historical Address by Elihu Root, June 17, 1912," in *Documentary History of Hamilton College*, ed. Joseph D. Ibbotson and S. N. D. North (Clinton, N.Y.: Hamilton College, 1922), 1–22.

36  See Hamilton College's Washington Alumni Association, "Minute on the Death of Dr. Oren Root," *Hamilton Record* 7 (1908): 90–92; Hamilton College, *Hamilton College Catalogue of the Officers and Students 1865–66* (Utica, N.Y.: Roberts, 1865); and Hamilton College, *Hamilton College Catalogue of the Officers and Students 1869–70* (Utica, N.Y.: Roberts, 1869).

37  Elias Loomis, *Elements of Analytical Geometry and of the Differential and Integral Calculus* (New York: Harper & Brothers, 1851), iii.

38  H. A. Newton, *A Memoir of Elias Loomis* (Washington, D.C.: Government Printing Office, 1891).

39  Loomis, *Analytical Geometry*, iii.

40  "Burial of Calculus by the Class of '59, Nulli Secundus, Hamilton College, Friday, Dec. 4, 1857," collection 1 0008 15016, Hamilton College Archives, Hamilton College.

41  "Burial of Calculus."

42  "Burial of Calculus."

43  Nash, Mireles, and Scott-Williams, "'Mattie Matix,'" 61–62.

44  Garber, *Vested Interests*.

45  William Smyth, *Elements of Algebra* (Portland, Maine: Shirley & Hyde, 1830).

46  William Smyth, *Elements of the Differential and Integral Calculus* (Brunswick, Maine: J. Griffin, 1854), 13–15.

47  "Burial of Mathematics by the Junior Class of Bowdoin College, August 30, 1853," folder 10.2, Special Collections and Archives, Bowdoin College.

48  "Burial of Mathematics."

49  "Burial of Mathematics."

50  "Burial of Mathematics."

51  "Burial of Euclid, November 1851, Programme of Exercises," folder 10.2, Special Collections and Archives, Bowdoin College; and [Hall], *College Words and Customs*, 45–46.

52  "Burial of Mathematics."

53  James McLachlan, "The *Choice of Hercules*: American Student Societies in the Early 19th Century," in *The University in Society*, ed. Lawrence Stone (Princeton, N.J.: Princeton University Press, 1974), 2:449–473; Thomas S. Harding, *College Literary Societies: Their Contribution to Higher Education in the United States, 1815–1876* (New York: Pageant Press Internal, 1971); and Roger L. Geiger, "Introduction: New Themes in the History of Nineteenth-Century Colleges," in *The American College in the Nineteenth Century*, ed. by Roger L. Geiger (Nashville: Vanderbilt University Press, 2000), 1–36.

54 "Jesus Loves Me Hymn Tune," *Hymnary*, last modified 2018, http://www.hymnary
.org/tune/jesus_loves_me_bradbury.

55 "60 Order of Exercises, at Calculus, His Burning, July 26, 1859," folder 10.2, Special
Collections and Archives, Bowdoin College.

56 Bowdoin College, *General Catalogue of Bowdoin College and the Medical School of
Maine, 1794–1894* (Brunswick, Maine: Bowdoin College Press, 1894), 62–63.

57 Horowitz, *Campus Life*, 3–22.

58 Bates College, "150 Years," *Bates 150 Years*, last modified October 2011, http://www
.bates.edu/150-years/.

59 [Hall], *College Words and Customs*, 27–32.

60 Fiss, "Mathematics and Mourning."

61 G. Tillson to P. S. Wilder, October 12, 1931, in folder 10.2, Special Collections and
Archives, Bowdoin College.

62 General Records, CA 05.20 Class Records, Muskie Archives, Bates College.

63 Bates College, "Chapter 2," *Bates 150 Years*, last modified October 2011, http://www
.bates.edu/150-years/.

64 Michael David Cohen, *Reconstructing the Campus: Higher Education and the
American Civil War* (Charlottesville: University of Virginia Press, 2012).

65 Winstead, *When Colleges Sang*.

66 Fiss, "Mathematics and Mourning."

67 General Records, CA 05.20 Class Records, Muskie Archives, Bates College.

68 *American College Song Book: A Collection of the Songs of Fifty Representative American Colleges* (Boston: Oliver Ditson, 1882); and *College Songs: A Collection of the
Most Popular Songs of the Colleges of America* (Boston: Oliver Ditson, 1882).

69 *Bates Student* newspaper articles from 1879, 1881, 1885, 1888, 1890, 1892, and 1893,
Muskie Archives, Bates College.

70 "Owldom," *Bates Student* 21, no. 10 (December 1893): 275.

71 Bates People Files, "Nichols, Charles," Muskie Archives, Bates College.

72 Bates People Files.

73 *Bates Student* newspaper articles from 1879, 1881, 1885, 1888, 1890, 1892, and 1893,
Muskie Archives, Bates College.

## Chapter 3   How Math Anxiety Has Developed from Classroom Tech

1 Michael J. Barany and Donald MacKenzie, "Chalk: Materials and Concepts in
Mathematics Research," in *Representation in Scientific Practice Revisited*, ed. Cateli-
jne Coopmans, Janet Vertesi, Michael E. Lynch, and Steve Woolgar (Cambridge,
Mass.: MIT Press, 2014), 107–130.

2 Peggy Aldrich Kidwell, Amy Ackerberg-Hastings, and David Lindsay Roberts,
*Tools of American Mathematics Teaching, 1800–2000* (Washington, D.C.: Smithsonian Institution / Baltimore, Md.: Johns Hopkins University Press, 2008), 21–34.

3 Kidwell, Ackerberg-Hastings, and Roberts, 21–34.

4 See definition 4 in *OED Online*, s.v. "anxiety, n.," last modified March 2019, https://
www.oed.com/view/Entry/8968.

5 See especially Sheila Tobias, *Overcoming Math Anxiety* (New York: Norton, 1978);
and Sheila Tobias, *Succeed with Math: Every Student's Guide to Conquering Math
Anxiety* (New York: College Entrance Examination Board, 1987).

6 See etymology in *OED Online*, s.v. "fright, n.," last modified March 2019, https://
www.oed.com/view/Entry/74683.

7 Sara Solovitch, *Playing Scared: A History and Memoir of Stage Fright* (New York: Bloomsbury, 2015); and Scott Stossel, "Performance Anxiety in Great Performers: What Hugh Grant, Gandhi, and Thomas Jefferson Have in Common," *Atlantic*, January/February 2014.

8 Educational perspectives are mainly about music, though sometimes about speech: Horst Hildebrandt, Matthias Nübling, and Victor Candia, "Increment of Fatigue, Depression, and Stage Fright during the First Year of High-Level Education in Music Students," *Medical Problems of Performing Artists* 27, no. 1 (March 2012): 43–48; Julie Jaffee Nagel, *Managing Stage Fright: A Guide for Musicians and Music Students* (Oxford: Oxford University Press, 2017); Liliana S. Araújo et al., "Fit to Perform: An Investigation of Higher Education Music Students' Perceptions, Attitudes, and Behaviors toward Health," *Frontiers in Psychology* 8 (2017): 1558; and the classic Theodore Clevenger Jr., "A Synthesis of Experimental Research in Stage Fright," *Quarterly Journal of Speech* 45, no. 2 (1959): 134–145.

9 Alan G. Gross and Joseph E. Harmon, *Science from Sight to Insight: How Scientists Illustrate Meaning* (Chicago: University of Chicago Press, 2014), 231–265; and Mike Markel and Stuart Selber, *Technical Communication*, 12th ed. (Boston: Bedford / St. Martin's, 2019), 580–610.

10 Christopher J. Phillips, "An Officer and a Scholar: Nineteenth-Century West Point and the Invention of the Blackboard," *History of Education Quarterly* 55 (2015): 82–108.

11 See Amy Ackerberg-Hastings, "Mathematics Is a Gentleman's Art: Analysis and Synthesis in American College Geometry Teaching, 1790–1840" (PhD diss., Iowa State University, 2000); V. Frederick Rickey and Amy Shell-Gellasch, "201 Years of Mathematics at West Point," in *West Point: Two Centuries and Beyond*, ed. Lance Betros (College Station: Texas A&M University Press, 2004), 586–613; and Phillips, "Officer and Scholar," 87–88.

12 Alex Roland, "Science and War," in *Historical Writing on American Science: Perspectives and Prospects*, ed. Sally Gregory Kohlstedt and Margaret W. Rossiter (Baltimore, Md.: Johns Hopkins University Press, 1985), 251–252; and Oliver Wolcott to James McHenry, 18 July 1800, as quoted in *Memoirs of the Administrations of Washington and John Adams* (New York: Printed for the Subscribers [W. Van Norden, Printer], 1846), 2:381–382.

13 See George Fielding, *Sylvanus Thayer of West Point* (New York: Julian Messner, 1959); Red Reeder, *Heroes and Leaders of West Point* (New York: T. Nelson, 1970), 19–21; James William Kershner, *Sylvanus Thayer: A Biography* (New York: Arno, 1982); Stephen E. Ambrose, *Duty, Honor, Country: A History of West Point* (Baltimore, Md.: Johns Hopkins University Press, 1999), 62–105; and Theodore J. Crackel, *West Point: A Bicentennial History* (Lawrence: University Press of Kansas, 2002).

14 Phillips, "Officer and Scholar," 87–88.

15 John H. B. Latrobe, *West Point Reminiscences from September, 1818, to March, 1882* (East Saginaw, Mich.: Evening News, 1887), 30.

16 Kidwell, Ackerberg-Hastings, and Roberts, *American Mathematics Teaching*, 22–23.

17 John Bogart, *The John Bogart Letters* (New Brunswick, N.J.: Rutgers College, 1914), 18.

18 Emmor Kimber, *Arithmetic Made Easy for Children* (Philadelphia: Kimber and Conard, 1809), iv.

19 Quoted in Florian Cajori, *The Teaching and History of Mathematics in the United States* (Washington, D.C.: Government Printing Office, 1890), 117.

20 Cajori, 116.

21 Latrobe, *West Point Reminiscences*, 9.

22 A variant of the translation in David Eugene Smith, *A Source Book in Mathematics* (New York: Dover, 1959), 326.

23 See Cajori, *Teaching and History of Mathematics*, 116–118; and John C. Greene, *American Science in the Age of Jefferson* (Ames: Iowa State University Press, 1984), 131.

24 Latrobe, *West Point Reminiscences*, 29.

25 Latrobe, 29.

26 Latrobe, 29.

27 Kidwell, Ackerberg-Hastings, and Roberts, *American Mathematics Teaching*, 24–25.

28 Joseph Henry, "Journal of a Trip to West Point and New York," *Joseph Henry Papers* (Washington, D.C.: Smithsonian Institution, 1972), 1:157.

29 Latrobe, *West Point Reminiscences*, 29.

30 Charles Davies, *Elements of Descriptive Geometry: With Their Application to Spherical Trigonometry, Spherical Projections, and Warped Surfaces* (Philadelphia: H. C. Carey and I. Lea, 1826), iv.

31 Kidwell, Ackerberg-Hastings, and Roberts, *American Mathematics Teaching*, 3–20.

32 Phillips, "Officer and Scholar."

33 Terry S. Reynolds, "The Education of Engineers in America before the Morrill Act of 1862," *History of Education Quarterly* 32 (1992): 459–482.

34 See Trinity College, *Catalogus Senatus Academici Collegii Santissimae Trinitatis Harfordiae in Republica Connecticutensi* (Hartford, Conn.: Trinity College, 1890), 10–16; and Glenn Weaver, *History of Trinity College* (Hartford, Conn.: Trinity College, 1967), 3–27.

35 Charles Davies, *Elements of Analytical Geometry: Embracing the Equations of the Point, the Straight Line, the Conic Sections, and Surfaces of the First and Second Order* (New York: Wiley & Long, 1836), 6.

36 Trinity College Archives, "Trinity Traditions," Trinity College, content copyright 2014, http://www.trincoll.edu/LITC/Watkinson/archives/Pages/traditions.aspx.

37 Caroline Winterer, *The Culture of Classicism: Ancient Greece and Rome in American Intellectual Life, 1780–1910* (Baltimore, Md.: Johns Hopkins University Press, 2002), 77–98.

38 Charles Davies, *Analytical Geometry* (Hartford, Conn.: A. S. Barnes, 1839).

39 Charles Davies, *The Logic and Utility of Mathematics, with the Best Methods of Instruction Explained and Illustrated* (New York: A. S. Barnes & Burr, 1860).

40 Davies, 293–340. Chapter 1 (293–307) concerns mental discipline. Chapter 2 (308–324) is about knowledge acquisition. Chapter 3 (325–340) is about the "practical" usefulness of mathematics.

41 Kidwell, Ackerberg-Hastings, and Roberts, *American Mathematics Teaching*, 25.

42 Quoted in Winterer, *Culture of Classicism*, 44–49.

43 Winterer, 48–49.

44 John Locke, *Posthumous Works of Mr. John Locke* (London: Printed by W. B. for A. and J. Churchill at the Black Swan in Pater-Noster-Row, 1706), 17.

45 *Reports on the Course of Instruction in Yale College: By a Committee of the Corporation, and the Academical Faculty* (New Haven, Conn.: Hezekiah Howe, 1828), 32–33.

46 *Instruction in Yale College*, 34–35.

47  See R. Freeman Butts, *The College Charts Its Course: Historical Conceptions and Current Proposals* (New York: McGraw-Hill, 1939), 118–125; Melvin I. Urofsky, "Reforms and Response: The Yale Report of 1828," *History of Education Quarterly* 5 (1965): 53–67; Brooks M. Kelley, *Yale: A History* (New Haven, Conn.: Yale University Press, 1974), 161; Stanley Guralnick, *Science and the Ante-bellum American College* (Philadelphia: American Philosophical Society, 1975), 28–33; Jack C. Lane, "The Yale Report of 1828 and Liberal Education: A Neo-republican Manifesto," *History of Education Quarterly* 27 (1987): 325–338; Frederick Rudolph, *The American College and University: A History* (Athens: University of Georgia Press, 1991), 130–135; David B. Potts, "Curriculum and Enrollment: Assessing the Popularity of Antebellum Colleges," in *The American College in the Nineteenth Century*, ed. Roger L. Geiger (Nashville: Vanderbilt University Press, 2000), 39–40; Winterer, *Culture of Classicism*, 48–49; and Jurgen Herbst, "The Yale Report of 1828," *International Journal of the Classical Tradition* 11 (2004): 213–231.

48  Daniel Coit Gilman, Harry Thurston Peck, and Frank Moore Colby, eds., *The New International Encyclopedia* (New York: Dodd, Mead, 1906), s.v. "Silliman, Benjamin."

49  Jeremiah Day and James Luce Kingsley, "Original Papers in Relation to a Course of Liberal Education," *American Journal of Science and Arts* 15 (1829): 297–351.

50  Quoted in the letter to Mr. John Horton, Goshen, 7 August 1830, folder YRG 41-C 1–8, "Conic Rebellion 1830: Statements and Drafts of Statements 1830," Yale University Archives.

51  Sean Wilentz, *The Rise of American Democracy: Jefferson to Lincoln* (New York: W. W. Norton, 2005).

52  See William Lathrop Kingsley, *Yale College: A Sketch of Its History, with Notices of Its Several Departments, Instructors, and Benefactors, Together with Some Account of Student Life and Amusements* (New York: H. Holt, 1879), 1:137–138; Clarence Deming, "Yale Wars of the Conic Sections," *Independent* 56 (1904): 667–669; and Kelley, *Yale*, 167–168.

53  Petition to the Faculty of Yale College, n.d. [29 July 1830], folder YRG 41-C 1–8, "Conic Rebellion 1830: Statements and Drafts of Statements 1830," Yale University Archives.

54  Petition, n.d. [30 July 1830], folder YRG 41-C 1–8, "Conic Rebellion 1830: Statements and Drafts of Statements 1830," Yale University Archives.

55  Letters, n.d. [30 July 1830], folder YRG 41-C 1–8, "Conic Rebellion 1830: Statements and Drafts of Statements 1830," Yale University Archives.

56  "At a Meeting of the Faculty of Yale College," 31 July 1830, folder YRG 41-C 1–8, "Conic Rebellion 1830: Statements and Drafts of Statements 1830," Yale University Archives.

57  Letter [to Jeremiah Day], 31 July 1830, folder YRG 41-C 1–8, "Conic Rebellion 1830: Statements and Drafts of Statements 1830," Yale University Archives.

58  Letters to Jeremiah Day, August 1830, folder YRG 41-C 1–10, Conic Rebellion 1830: Correspondence 1830–31, Yale University Archives.

59  Josiah Quincy to Jeremiah Day, 24 August 1830, folder YRG 41-C 1–10, Conic Rebellion 1830: Correspondence 1830–31, Yale University Archives.

60  Letter from Yale, 3 August 1831, folder YRG 41-C 1–10, Conic Rebellion 1830: Correspondence 1830–31, Yale University Archives.

61  Kelley, *Yale*, 168–169.

62 *A Circular, Explanatory of the Recent Proceedings of the Sophomore Class, in Yale College* (New Haven, 1830).

63 Kimber, *Arithmetic Made Easy*, iv.

64 [Lyman Hotchkiss Bagg], *Four Years at Yale* (New Haven, Conn.: Charles C. Chatfield, 1871), 324–325.

65 [B. H. Hall], *A Collection of College Words and Customs* (Cambridge, Mass.: John Bartlett, 1851), 27–31.

66 *Yale Literary Magazine* 24, no. 7 (June 1859): 325.

67 W. H. Davenport (del.), lithograph by Emil Crisand, New Haven, Conn., 1858, personal collection.

68 [Bagg], *Four Years*, 319–326.

69 [Bagg], 324–325. Also see George Henry Nettleton, ed., *The Book of the Yale Pageant* (New Haven, Conn.: Yale University Press, 1916), 80–81.

70 See Thomas D. Snyder, ed., *120 Years of American Education: A Statistical Portrait* (Washington, D.C.: National Center for Education Statistics, 1993), 76.

## Chapter 4    How Math Communication Has Been Theatrical

1 W. H. Davenport (del.), lithograph by Emil Crisand, New Haven, Conn., 1858, personal collection.

2 Kim Tolley, *The Science Education of American Girls: A Historical Perspective* (London: RoutledgeFalmer, 2003), 75–94; John L. Rury, *Education and Women's Work: Female Schooling and the Division in Urban America, 1870–1930* (Albany: State University of New York Press, 1991), 11–48, 131–174; William J. Reese, *The Origins of the American High School* (New Haven, Conn.: Yale University Press, 1995), 208–235; and Ellen Condliffe Lagemann, *An Elusive Science: The Troubling History of Education Research* (Chicago: University of Chicago Press, 2000), 1–18.

3 Lynn D. Gordon, *Gender and Higher Education in the Progressive Era* (New Haven, Conn.: Yale University Press, 1990), 52–84; and David M. Stameshkin, *The Town's College: Middlebury College, 1800–1915* (Middlebury, Vt.: Middlebury College, 1985), 193–227.

4 Laurie A. Wilkie, *The Lost Boys of Zeta Psi: A Historical Archaeology of Masculinity at a University Fraternity* (Berkeley: University of California Press, 2010), 123–154.

5 Sara Hottinger, *Inventing the Mathematician: Gender, Race, and Our Cultural Understandings of Mathematics* (Albany, N.Y.: SUNY Press, 2016), 1–2.

6 Heather Mendick, "A Beautiful Myth? The Gendering of Being/Doing 'Good at Maths,'" *Gender and Education* 17, no. 2 (2005): 203–219; Heather Mendick, "Mathematical Stories: Why Do More Boys Than Girls Choose to Study Mathematics at AS-Level in England?," *British Journal of Sociology of Education* 26, no. 2 (2005): 235–251; and Valerie Walkerdine, *Counting Girls Out* (London: Routledge, 1998).

7 See Andrew Fiss, "Studying Objects, Objectifying Students: Natural History at Women's Colleges in New York State," *New York History* 98, no. 2 (2017): 205–229.

8 About the heavily documented "South Carolina Cadets," see Maximilian LaBorde, *History of the South Carolina College: From Its Incorporation, Dec. 19, 1801, to Dec. 19, 1865* (Charleston, S.C.: Walker, Evans, and Cogswell, 1874), 455–460; Irdell Jones, "The South Carolina College Cadets," in *A History of the University of South Carolina*, ed. Edwin L. Green (Columbia, S.C.: State Company, 1916), 361–388; and Daniel Walker Hollis, *University of South Carolina: College to University*

(Columbia: University of South Carolina Press, 1956), 213–220. For other institutional analyses, see Herbert B. Adams, *The College of William and Mary: A Contribution to the History of Higher Education* (Washington, D.C.: Government Printing Office, 1887), 61; and Michael David Cohen, *Reconstructing the Campus: Higher Education and the American Civil War* (Charlottesville: University of Virginia Press, 2012), 128–152.

9  Mark Twain and Charles Dudley Warner, *The Gilded Age: A Tale of Today* (Chicago: American Publishing, 1873). For a pithy review of the term's use in twenty-first-century historical scholarship, see Elisabeth Israels Perry, "Men Are from the Gilded Age, Women Are from the Progressive Era," *Journal of the Gilded Age and Progressive Era* 1, no. 1 (January 2002): 25–48.

10  About educational philanthropy (emphasizing the case of Cornell), see Laurence R. Veysey, *The Emergence of the American University* (Chicago: University of Chicago Press, 1965), 81–98. Also about Cornell, see W. P. Rogers, *Andrew D. White and the Modern University* (Ithaca, N.Y.: Cornell University Press, 1942); C. L. Becker, *Cornell University: Founders and the Founding* (Ithaca, N.Y.: Cornell University Press, 1943); and Morris Bishop, *A History of Cornell* (Ithaca, N.Y.: Cornell University Press, 1962). About Vassar, see Maryann Bruno and Elizabeth A. Daniels, *Vassar* (Charleston, S.C.: Arcadia, 2001), 19; and Carl N. Degler, "Vassar College," in *American Places: Encounters with History*, ed. William E. Leuchtenburg (Oxford: Oxford University Press, 2000), 93–104. About Wells, see Jane Marsh Dieckmann, *Wells College: A History* (Aurora, N.Y.: Wells College Press, 1995).

11  Thomas Woody, *A History of Women's Education in the United States* (New York: Science Press, 1929); Tolley, *Science Education of American Girls*; and Andrea G. Radke-Moss, *Bright Epoch: Women and Coeducation in the American West* (Lincoln: University of Nebraska Press, 2008).

12  J. M. Taylor, *Before Vassar Opened: A Contribution to the History of the Higher Education of Women in America* (Boston: Houghton Mifflin, 1914).

13  See Tolley, *Science Education of American Girls*, 83–84; and Andrew Fiss, "Cultivating Parabolas in the Parlor Garden: Reconciling Mathematics Education and Feminine Ideals in Nineteenth-Century America," *Science & Education* 23, no. 1 (2014).

14  Andrew Fiss, "Mathematics and Mourning: Textbook Burial and Student Culture before and after the Civil War, 1853–1880," *History of Education Quarterly* 57, no. 2 (2017): 221–246.

15  James Lloyd Winstead, *When Colleges Sang: The Story of Singing in American College Life* (Birmingham: University of Alabama Press, 2013), 116–152.

16  *Acta Columbiana*, 9 April 1880. Also quoted in Michael Rosenthal, *Nicholas Miraculous: The Amazing Career of the Redoubtable Dr. Nicholas Murray Butler* (New York: Columbia University Press, 2015), 44.

17  Tolley, *Science Education of American Girls*, 13–34.

18  See Erik Kogut, "A Columbia Moment: Burial of the Ancient," *Bwog: Columbia Student News*, 15 October 2010, https://bwog.com/2010/10/a-columbia-moment-burial-of-the-ancient/; and *WikiCU, the Columbia University wiki encyclopedia*, s.v. "Burial of the Ancient," last modified 1 December 2013, http://www.wikicu.com/Burial_of_the_Ancient.

19  E. F. Bojesen, *A Manual of Grecian and Roman Antiquities* (New York: Appleton, 1848).

20  Bojesen, iii, v–vi.

21  Bojesen, 66–72, 140–143, 150–151.

22 Kogut, "Columbia Moment."
23 About the move, see Robert McCaughey, *Stand, Columbia: A History of Columbia University in the City of New York* (New York: Columbia University Press, 2003), 116–143; and John Howard Van Amringe, *A History of Columbia University, 1754–1904* (New York: Columbia University Press, 1904), 123–139.
24 On the importance of geography and the route to class, see Richard C. Sadler and Don J. Lafreniere, "You Are Where You Live: Methodological Challenges to Measuring Children's Exposure to Hazards," *Journal of Children and Poverty* 23, no. 2 (2017): 189–198.
25 Rutgers Female College appears in Helen Lefkowitz Horowitz, "Roar, Alma Mater, Roar," *Reviews in American History* 34, no. 1 (2006): 81–85.
26 See Rutgers Female College, *Proceedings of the Meeting Held at the Inauguration of Rutgers Female College, April 26, 1867* (New York: Agathynian, 1867); and Fiss, "Studying Objects," 205–229.
27 Similar cases appear in Winstead, *When Colleges Sang*, 134.
28 Winstead, 22–23.
29 "The Burial of the Ancient," *Columbia Daily Spectator*, 1 June 1879; and "The Triumph," *Columbia Daily Spectator*, 15 June 1883.
30 "Burial of the Ancient."
31 "About College," *Columbia Daily Spectator*, 10 February 1880. Also, see Rosenthal, *Nicholas Miraculous*, 44.
32 Lewis Sayre Burchard, "The Later Seventies," in *The City College: Memories of Sixty Years*, ed. Philip J. Mosenthal and Charles F. Horne (New York: G. P. Putnam's Sons, 1907), 287.
33 Burchard.
34 See "Excursion and Cremation," *Cornell Daily Sun*, 19 May 1887; and "Excursion and Cremation," *Cornell Daily Sun*, 9 May 1888.
35 "O.W.J. in Ashes," *Cornell Era* (Ithaca, N.Y.: Andrus & Church, 1882), 291–292.
36 "O.W.J. in Ashes," 291–292.
37 Bishop, *History of Cornell*, 121. Also see Rogers, *Andrew D. White*; and Becker, *Cornell University*.
38 Andrew Dickson White, *My Reminiscences of Ezra Cornell* (Ithaca, N.Y.: Cornell University Press, 1890), 9. Quoted extensively, as in Richard H. Penner, *Cornell University* (Charleston, S.C.: Arcadia, 2013), 77. On traditions of its quoting, see Cohen, *Reconstructing the Campus*, 87.
39 Florian Cajori, *The Teaching and History of Mathematics in the United States* (Washington, D.C.: Government Printing Office, 1890), 180–186; and Karen Hunger Parshall and David E. Rowe, *The Emergence of the American Mathematical Research Community, 1876–1900: J. J. Sylvester, Felix Klein, and E. H. Moore* (Providence, R.I.: American Mathematical Society, 1994), 270.
40 Bishop, *History of Cornell*, 86–87, 167.
41 Profs. Oliver, Wait, and Jones, *A Treatise on Algebra* (Ithaca, N.Y.: Ginn, Heath, 1882). Also, see "O.W.J. in Ashes," 291–292.
42 Jeremiah Day, *An Introduction to Algebra, Being the First Part of a Course of Mathematics, Adapted to the Method of Instruction in the American Colleges* (New Haven, Conn.: Howe & Deforest, 1814), 5.
43 Oliver, Wait, and Jones, *Algebra*, 6, 10.
44 Dieckmann, *Wells College*.

45 Quoted in Judith Lavelle, "Wells College History," *Wells College*, https://www.wells.edu/about/wells-college-history.

46 See Bishop, *History of Cornell*, 174; and Dieckmann, Wells College.

47 "Sophomore Excursion," *Cornell Era* (Ithaca, N.Y.: Andrus & Church, 1882), 340.

48 "Sophomore Excursion," 340.

49 "Sophomore Excursion," 341.

50 "Sophomore Excursion," 341.

51 Winstead, *When Colleges Sang*, 134.

52 Among many sources, see Nancy Woloch, *Women and the American Experience* (Boston: McGraw-Hill, 2000), 119–125.

53 "V.C. Trig Ceremonies," subject file 25.35, Archives and Special Collections Library, Vassar College. Also, see Winstead, *When Colleges Sang*, 114–115, 209, 230.

54 *Trial of Trigonometry at Vassar College, February 7th, 1871* (Poughkeepsie: Dutchess Farmer Steam Print, 1871), in "V.C. Trig Ceremonies," subject file 25.35, Archives and Special Collections Library, Vassar College.

55 *Trial of Trigonometry*, 4.

56 *Trial of Trigonometry*, 7.

57 *Trial of Trigonometry*.

58 Matthew Vassar, *Communications to the Board of Trustees of Vassar College* (New York: John A. Gray and Green, 1869), 5–8.

59 University of the State of New York, *Eighty-Second Annual Report of the Regents of the University, Made to the Legislature February 26, 1869* (Albany, N.Y.: Argus, 1869).

60 Among the many instances of the term *experiment*, see the quoted remarks of Vassar College president John H. Raymond: James Monroe Taylor and Elizabeth Hazelton Haight, *Vassar* (New York: Oxford University Press, 1915), 77.

61 Helen Lefkowitz Horowitz, *Alma Mater: Design and Experience in the Women's Colleges from Their Nineteenth-Century Beginnings to the 1930s* (Amherst: University of Massachusetts Press, 1993), 38–40.

62 Sebastian Langdell, "Overview of Original Faculty," *Vassar Encyclopedia*, last modified 2004, http://vcencyclopedia.vassar.edu/faculty/original-faculty/original-faculty.html.

63 Archives folder 4.80: Natural History, Archives and Special Collections Library, Vassar College.

64 The fate of the sign became a favorite joke. A member of the class of 1892 wrote a parodic song about how "A strong east wind" scandalously "tore the 'Female' off . . . the College." See Amy L. Reed, "Vassar College," in *Vassar College: Seventy-Fifth Anniversary Song Book* (Poughkeepsie: Vassar College Press, 1940), 7–8.

65 Mary Harriott Norris, *The Golden Age of Vassar* (Poughkeepsie: Vassar College Press, 1915), 65.

66 Frances A. Wood, *Earliest Years at Vassar: Personal Recollections* (Poughkeepsie: Vassar College Press, 1909), 29.

67 Ellen Swallow, 25 November 1868, Biographical File for "Richards, Ellen Swallow," folder 3 (Richards Misc Letters, 1869–1903), Archives and Special Collections Library, Vassar College Libraries. Emphasis added.

68 *Trial of Trigonometry*, 7, 11, 13.

69 *Trial of Trigonometry*.

70 *Trial of Trigonometry*, 13.

71  Edward H. Clarke, *Sex in Education; Or, A Fair Chance for the Girls* (Boston: James R. Osgood, 1873). Clarke's book can be found in a variety of libraries, including Rare Books Collection s.v. "Clarke, Edward H.," number 1.Mh.1873.C, Countway Library of Medicine, Harvard University, Cambridge, Massachusetts.

72  For a small cross section, see Mary Roth Walsh, *"Doctors Wanted: No Women Need Apply": Sexual Barriers in the Medical Profession, 1835–1975* (New Haven, Conn.: Yale University Press, 1977), 120–125; Vern Bullough and Martha Voght, "Women, Menstruation, and Nineteenth-Century Medicine," *Bulletin of the History of Medicine* 47, no. 1 (1973): 66–82; Rosalind Rosenberg, *Beyond Separate Spheres: Intellectual Roots of Modern Feminism* (New Haven, Conn.: Yale University Press, 1983), chap. 1; Sue Zschoche, "Dr. Clarke Revisited: Science, True Womanhood, and Female Collegiate Education," *History of Education Quarterly* 29, no. 4 (1989): 545–569; Sharra Vostral, *Under Wraps: A History of Menstrual Hygiene Technology* (Lanham, Md.: Lexington Books, 2008), 26–35; Carla Bittel, *Mary Putnam Jacobi and the Politics of Medicine in Nineteenth-Century America* (Chapel Hill: University of North Carolina Press, 2009), chap. 4; and Kimberly A. Hamlin, *From Eve to Evolution: Darwin, Science, and Women's Rights in Gilded Age America* (Chicago: University of Chicago Press, 2014), 73–80.

73  Clarke, *Sex in Education*, 83.

74  "Home Matters," *Vassar Miscellany*, March 1, 1881.

75  "Home Matters."

76  "Home Matters."

77  Andrew Fiss and Laura Kasson Fiss, "Laughing Out of Math Class: The Vassar *Mathematikado* and Nineteenth-Century Women's Education," *Configurations* 27, no. 3 (2019): 301–329.

78  "V.C. Trig Ceremonies," subject file 25.35, Archives and Special Collections Library, Vassar College.

79  G. A. Wentworth, *Elements of Plane and Solid Geometry* (Boston: Ginn and Heath, 1880), v–vi.

80  W. E. Byerly, *Chauvenet's Treatise on Elementary Geometry* (Philadelphia: J. B. Lippincott, 1892 [1887]), 253.

81  About Vassar, see Marion Bacon, *Life at Vassar: Seventy-Five Years in Pictures, 1865–1940* (Poughkeepsie: Vassar Cooperative Bookshop, 1940), 92–95. Yopie Prins discusses the costumes worn at Smith College, Bryn Mawr College, and Cambridge's Girton College in Yopie Prins, *Ladies' Greek: Victorian Translations of Tragedy* (Princeton: Princeton University Press, 2017), 95–115, 116–151, 218–232.

82  See references to the Wellesley Shakespeare Society, student plays at Girton, and the later Radcliffe Idlers in Barbara Miller Solomon, *In the Company of Educated Women: A History of Women and Higher Education in America* (New Haven, Conn.: Yale University Press, 1985), 105–106, 129.

83  "The Mathematikado: The Mystical 'Trig Ceremonies' of Sweet Girl Graduates," *Cornell Daily Sun*, 7 April 1886.

84  "The Mathematikado," *Rome Daily Sentinel*, March 1886; and "The Mathematikado," *New York Times*, 24 March 1886.

85  "Exchanges," *Stevens Indicator*, April 1886; and "The Mathematikado: 'Trig Ceremonies' at Vassar," *Harvard Crimson*, 25 March 1886.

86  Grace W. Soper, "Festival Days at Girls' Colleges," *St. Nicholas: An Illustrated Magazine for Young Folks* 20, no. 2 (1893): 682–691.

87  Soper, 687.

88  Soper, 682, 685.

89  Soper, 691.

90  Helen Lee Sherwood, "Informal Dramatics," in *The Vassar Miscellany: Vassar, 1865–1915, from the Undergraduate Point of View, Fiftieth Anniversary Number* (Poughkeepsie, N.Y.: Vassar College, 2015), 111.

91  Sherwood, 111.

92  Elma G. Martin diary, 1892, Student Diaries Collection, Archives and Special Collections Library, Vassar College.

93  Jean Webster, *When Patty Went to College* (New York: Century, 1903), chap. 13; Charles E. Bolton, *The Harris-Ingram Experiment* (Cleveland: Burrow Brothers, 1905), chap. 10; and Julia A. Schwartz, *Elinor's College Career* (Boston: Little, Brown, 1906), 95–101.

94  Alida C. Avery, *Sex and Education: A Reply to Dr. E. H. Clarke's "Sex in Education,"* ed. Julia Ward Howe (Boston: Roberts Brothers, 1874), 191–195; and James Orton, "Four Years in Vassar College," in *Addresses and Journal of Proceedings of the National Educational Association, Session of the Year 1874, at Detroit, Michigan* (Worcester, Mass.: Charles Hamilton, 1874), 109–117.

95  Della Dumbaugh Fenster and Karen Hunger Parshall, "Women in the American Mathematical Research Community, 1891–1906," in *The History of Modern Mathematics* (Boston: Academic Press, 1994), 3:228–261.

96  Mentioned in John R. Cross, "Whispering Pines: Written in Stone," *Bowdoin Daily Sun*, September 2010, http://www.bowdoindailysun.com/2010/09/whispering-pines-written-in-stone/.

97  Margaret A. Nash, Danielle C. Mireles, and Amanda Scott-Williams, "'Mattie Matix' and Prodigal Princes: A Brief History of Drag on College Campuses from the Nineteenth Century to the 1940s," in *Rethinking Campus Life: New Perspectives on the History of College Students in the United States*, ed. Christine A. Ogren and Marc A. VanOverbeke (New York: Palgrave Macmillan, 2018), 61–90.

98  My analysis parallels Wilkie, *Lost Boys*, 230–235.

## Chapter 5   How Math Anxiety Became about Written Testing

1  Peggy Aldrich Kidwell, Amy Ackerberg-Hastings, and David Lindsay Roberts, *Tools of American Mathematics Teaching, 1800–2000* (Washington, D.C.: Smithsonian Institution / Baltimore, Md.: Johns Hopkins University Press, 2008), 21–22.

2  Kidwell, Ackerberg-Hastings, and Roberts, 35–52.

3  Kidwell, Ackerberg-Hastings, and Roberts, 35–52.

4  For just a few recent examples, see Michael J. Barany and Donald MacKenzie, "Chalk: Materials and Concepts in Mathematics Research," in *Representation in Scientific Practice Revisited*, ed. by Catelijne Coopmans, Janet Vertesi, Michael E. Lynch, and Steve Woolgar (Cambridge, Mass.: MIT Press, 2014), 107–130; Kidwell, Ackerberg-Hastings, and Roberts, *American Mathematics Teaching*; Christopher J. Phillips, *The New Math: A Political History* (Chicago: University of Chicago Press, 2015); and Christopher G. White, *Other Worlds: Spirituality and the Search for Invisible Dimensions* (Cambridge, Mass.: Harvard University Press, 2018).

5  Michael David Cohen, *Reconstructing the Campus: Higher Education and the American Civil War* (Charlottesville: University of Virginia Press, 2012), 1–18.

6  See Willis Rudy, *Building America's Schools and Colleges: The Federal Contribution* (Cranbury, N.J.: Cornwall, 2003), 20; Earle D. Ross, *Democracy's Colleges: The*

*Land-Grant Movement in the Formative Stage* (Ames: Iowa State University Press, 1942), 56–60; and Cohen, *Reconstructing the Campus*, 54–55.

7 See Donald R. Warren, *To Enforce Education: A History of the Founding Years of the United States Office of Education* (Detroit: Wayne State University Press, 1974); Williamjames Hull Hoffer, *To Enlarge the Machinery of Government: Congressional Debates and the Growth of the American State, 1858–1891* (Baltimore, Md.: Johns Hopkins University Press, 2007), 38–43; and Cohen, *Reconstructing the Campus*, 165–173.

8 See Margaret A. Nash, *Women's Education in the United States, 1780–1840* (New York: Palgrave Macmillan, 2005); Hilary J. Moss, *Schooling Citizens: The Struggle for African American Education in Antebellum America* (Chicago: University of Chicago Press, 2009); Carl F. Kaestle, *Pillars of the Republic: Common Schools and American Society, 1780–1960* (New York: Hill & Wang, 1983), 120–121; David Tyack and Elisabeth Hansot, *Learning Together: A History of Coeducation in American Schools* (New Haven, Conn.: Yale University Press, 1990), 121–122; and Cohen, *Reconstructing the Campus*, 6–9.

9 Deborah J. Warner, "Geography of Heaven and Earth, Part 4," *Rittenhouse* 2 (1988): 109–137.

10 See Asa Briggs, *Victorian Things* (London: B. T. Batsford, 1988), 182–187; Henry Petroski, *The Pencil: A History of Design and Circumstance* (New York: Knopf, 1990); David C. Smith, *History of Papermaking in the United States, 1691–1969* (New York: Lakewood, 1971), 121–187; Charnel Anderson, *Technology in American Education* (Washington, D.C.: Government Printing Office, 1962), 34–37; and Kidwell, Ackerberg-Hastings, and Roberts, *American Mathematics Teaching*, 35.

11 For example, Lauren Gunderson, *Silent Sky* (New York: Dramatis Play Service, 2015).

12 For brief mentions, see Phyllis Vine, "The Social Function of Eighteenth-Century Higher Education," *History of Education Quarterly* 16, no. 4 (1976): 409–424; and Arthur M. Cohen and Carrie B. Kisker, *The Shaping of American Higher Education: Emergence and Growth of the Contemporary System* (New York: John Wiley & Sons, 2010), 90.

13 Patricia Cline Cohen, *A Calculating People: The Spread of Numeracy in Early America* (Chicago: University of Chicago Press, 1982), 139–143.

14 Cohen, *Calculating People*, 139–143; and E. Jennifer Monaghan, *Learning to Read and Write in Colonial America* (Amherst: University of Massachusetts Press, 2005), 373. Also, see Andrew Fiss, "Cultivating Parabolas in the Parlor Garden: Reconciling Mathematics Education and Feminine Ideals in Nineteenth-Century America," *Science & Education* 23, no. 1 (2014): 241–250.

15 James Fishburn, *Some Remarks on Education, Textbooks Etc.* (Portland, Maine: Shirley and Hyde, 1828), 20.

16 Alma Lutz, *Emma Willard: Pioneer Educator of American Women* (Boston: Beacon, 1964), 76; Thomas Woody, *A History of Women's Education in the United States* (New York: Science Press, 1929), 309; and Mildred Sandison Fenner and Eleanor Craven Fishburn, *Pioneer American Educators* (Washington, D.C.: Hugh Birch-Horace Mann Fund of the National Education Association, 1944), 36.

17 Lutz, *Emma Willard*, 77; and Ilana DeBare, *Where Girls Come First: The Rise, Fall, and Surprising Revival of Girls' Schools* (New York: Penguin, 2004), 31.

18 Henry Fowler, "Educational Services of Mrs. Emma Willard," *American Journal of Education* 6 (1859): 125–168.

19  Albanian, "Waterford Female Academy," *Albany Gazette*, 23 January 1821, 3.

20  John Foster, *A Sketch of the Tour of General Lafayette* (Portland, Maine: A. W. Thayer, 1824), 170.

21  Foster, 171.

22  Quoted in Lucretia Maria Davidson, *Poetical Remains of the Late Lucretia Maria Davidson* (Philadelphia: Lea and Blanchard, 1841), 72–73.

23  Quoted in Davidson, 79.

24  Davidson, 71.

25  Davidson, 76.

26  Lucy Larcom, *A New England Girlhood, Outlined from Memory* (Cambridge, Mass.: Riverside Press, 1889), 266.

27  Phebe McKeen and Philena Fuller McKeen, *A History of Abbot Academy, Andover, Massachusetts, 1829–1879* (Andover, Mass.: Warren F. Draper, 1880), 18.

28  Nash, *Women's Education*, 87.

29  Nash, 87.

30  Kim Tolley, *The Science Education of American Girls: A Historical Perspective* (London: RoutledgeFalmer, 2003), 81.

31  Cohen, *Calculating People*, 146.

32  Fiss, "Cultivating Parabolas."

33  G. A. Dewitt and John Kingsbury, "Providence High School," repr. in the *American Journal of Education* 3 (1828): 428.

34  Dewitt and Kingsbury, 428.

35  Elias Marks, *Hints on Female Education* (Columbia, S.C.: David W. Sims, 1828), 19.

36  John T. Irving, *Address Delivered on the Opening of the New-York High-School for Females* (New York: William A. Mercein, 1826), 20–24.

37  Irving, 20–24.

38  Maria Budden, *Thoughts on Domestic Education* (London: Charles Knight, 1826), 59–65.

39  Almira Hart Lincoln Phelps, *The Female Student; Or, Lectures to Young Ladies on Female Education* (Boston: Crocker and Brewster, 1836), 98–99.

40  Reviewed in Fiss, "Cultivating Parabolas."

41  Lebbeus Booth, *Ballston Spa Female Seminary* (Albany, N.Y.: Packard and Van Benthuysen, 1824), 3–8.

42  Christopher J. Phillips, "An Officer and a Scholar: Nineteenth-Century West Point and the Invention of the Blackboard," *History of Education Quarterly* 55 (2015): 82–108.

43  See James M. McPherson, *Battle Cry of Freedom: The Civil War Era* (Oxford: Oxford University Press, 2003 [1988]), 308–338.

44  For a historical treatment of Confederate rank (in the case of Lee, in particular), see Edward D. C. Campbell, "The Fabric of Command: R. E. Lee, Confederate Insignia, and the Perception of Rank," *Virginia Magazine of History and Biography* 98 (April 1990): 261–290.

45  "(VII. Public—No. 57) An Act to Promote the Efficiency of the Corps of Engineers and of the Ordinance Department, and for Other Purposes," in *The War of the Rebellion: A Compilation of the Official Records of the Union and Confederate Armies* (Washington, D.C.: Government Printing Office, 1899), series 3, vol. 3: 93. Also, San Francisco's *Daily Evening Bulletin* had an article about the act in April 1863: "Public Act—No. 57: An Act to Promote the Efficiency of the Corps of Engineers and of the Ordnance Department, and for Other Purposes," *Daily Evening Bulletin*, 28 April 1863, no. 18, column E.

46 See Duane Schultz, *The Dahlgren Affair: Terror and Conspiracy in the Civil War* (New York: W. W. Norton, 1998); and Eric J. Wittenberg, *Like a Meteor Blazing Brightly: The Short but Controversial Life of Colonel Ulric Dahlgren* (Roseville, Minn.: Edinborough, 2009).

47 See Florian Cajori, *The Teaching and History of Mathematics in the United States* (Washington, D.C.: Government Printing Office, 1890); V. Frederick Rickey and Amy Shell-Gellasch, "201 Years of Mathematics at West Point," in *West Point: Two Centuries and Beyond*, ed. Lance Betros (College Station: Texas A&M University Press, 2004), 586–613; and V. Frederick Rickey and Amy Shell-Gellasch, "Mathematics Education at West Point: The First Hundred Years—Albert E. Church, Mathematics Professor, 1837–1878," *Convergence*, July 2010, https://www.maa.org/press/periodicals/convergence/mathematics-education-at-west-point-the-first-hundred-years-albert-e-church-mathematics-professor.

48 Rickey and Shell-Gellasch, "Mathematics Education."

49 "October 27, 1864" in Paul Dahlgren's diary, *Private Journal. No. 5*, folder 2238, Nineteenth Century Manuscripts Collection, Special Collections and Archives Division, United States Military Academy Library.

50 Albert E. Church commented on these historic divisions in the letter Albert E. Church to "My Dear General," 11 February 1873, Albert E. Church Vertical File, Special Collections and Archives Division, United States Military Academy Library. Also, biographies of certain "firsts" of West Point classes, such as Henry Eustis and Robert E. Lee, demonstrate their desires to be engineers. For Eustis, see Ezra J. Warner, "Henry Lawrence Eustis," in *Generals in Blue: Lives of the Union Commanders* (Baton Rouge: Louisiana State University Press, 1964), 144. For Lee, see Eben Smith, "The Military Education of Robert E. Lee," *Virginia Magazine of History and Biography* 35 (April 1927): 97–160; Charles Dudley Rhodes, *Robert E. Lee, the West Pointer* (Richmond: Garrett & Massie, 1932); John Morgan Dederer, "The Origins of Robert E. Lee's Bold Generalship: A Reinterpretation," *Military Affairs* 49 (July 1985): 117–123; and Michael Fellman, *The Making of Robert E. Lee* (New York: Random House, 2000).

51 Paul Dahlgren's diary, *Private Journal. No. 5*, folder 2238, Nineteenth Century Manuscripts Collection, Special Collections and Archives Division, United States Military Academy Library.

52 "November 7, 1864" in Paul Dahlgren's diary, *Private Journal. No. 5*, folder 2238, Nineteenth Century Manuscripts Collection, Special Collections and Archives Division, United States Military Academy Library.

53 "November 16, 1864" to "December 1, 1864," in Paul Dahlgren's diary, *Private Journal. No. 5*, folder 2238, Nineteenth Century Manuscripts Collection, Special Collections and Archives Division, United States Military Academy Library.

54 "December 1, 1864," in Paul Dahlgren's diary, *Private Journal. No. 5*, folder 2238, Nineteenth Century Manuscripts Collection, Special Collections and Archives Division, United States Military Academy Library.

55 See Ernest Dupuy, *Where They Have Trod: The West Point Tradition in American Life* (New York: Frederick A. Stokes, 1940), 163; and Phillips, "Officer and Scholar," 102n60.

56 Paul Dahlgren's diary, *Private Journal. No. 5*.

57 Paul Dahlgren's diary, *Private Journal. No. 5*.

58 Paul Dahlgren's diary, *Private Journal. No. 5*.

59 "March 4, 1865," in Paul Dahlgren's diary, *Private Journal. No. 5*, folder 2238, Nineteenth Century Manuscripts Collection, Special Collections and Archives Division, United States Military Academy Library.

60 Albert E. Church, *Elements of Descriptive Geometry, with Its Applications to Spherical Projections, Shades and Shadows, Perspective, and Isometric Projections* (New York: A. S. Barnes, 1864), ii.

61 Church, 1.

62 Church, 192.

63 "Descriptive on the Brain," in Paul Dahlgren's diary, *Private Journal. No. 5*, folder 2238, Nineteenth Century Manuscripts Collection, Special Collections and Archives Division, United States Military Academy Library.

64 For instance, see the depiction of epilepsy in Edwin Lee, *A Treatise on Some Nervous Disorders* (London: Burgess and Hill, 1833), 60; or *The Cyclopaedia; or, Universal Dictionary of Arts, Sciences and Literature*, vol. 12 (London: Logman, Hurst, Rees, Orme, and Brown, 1819), s.v. "Drunkenness."

65 Paul Dahlgren's diary, *Private Journal. No. 5*.

66 See Nancy Beadie, "From Student Markets to Credential Markets: The Creation of the Regents Examination System in New York State, 1864–1890," *History of Education Quarterly* 39, no. 1 (1999): 1–30.

67 Kidwell, Ackerberg-Hastings, and Roberts, *American Mathematics Teaching*, 37.

68 Kidwell, Ackerberg-Hastings, and Roberts, 37.

69 See the entry for "Phillips Academy, Exeter, N.H.," in "Eliot's Summary of Subject Offerings and Time Allotments in Forty High Schools," as quoted in Edward Krug, *The Shaping of the American High School* (New York: Harper & Row, 1964), 49.

70 See the entry for "Boston Public Latin School, Mass." in "Eliot's Summary," as quoted in Krug, *Shaping*, 51.

71 See the entry for "Battle Creek High School, Mich." in "Eliot's Summary," as quoted in Krug, *Shaping*, 48.

72 See G. Bailey Price, "Contributions of the 1893 Columbian Exposition to Mathematics," in *A Century of Mathematical Meetings*, ed. Bettye Anne Case (Providence, R.I.: American Mathematical Society, 1996), 49–60; and David Zitarelli, "Mathematics at World's Fairs: Chicago 1893 and St. Louis 1904" (paper, History of Science Society panel, "Studies in the Internationalization of Mathematics: Goals, Strategies, and the Outcomes in Nineteenth and Twentieth Centuries," Pittsburgh, Pa., 8 November 2008).

73 H. S. White, "A Brief Account of the Congress on Mathematics Held at Chicago in August, 1893," in *Mathematical Papers Read at the International Mathematical Congress Held in Connection with the World's Columbian Exposition Chicago 1893*, ed. E. H. Moore et al. (New York: Macmillan, 1896), vii–viii. Also quoted in Price, "Columbian Exposition to Mathematics," 54–55.

74 G. B. Halsted to Florian Cajori, 25 December 1888, as quoted in Cajori, *Teaching and History of Mathematics*, 265.

75 T. H. Safford, "Modern Mathematics in the College Course," in *The Addresses and Journal of Proceedings of the National Educational Association, Sessions of the Year 1871 at St. Louis, M.O.* (New York: James H. Holmes, 1872), 183–190.

76 See "Safford, Truman Henry," in *Appleton's Cyclopedia of American Biography*, ed. James Grant Wilson, John Fiske, and Stanley L. Klos (New York: Appleton, 1887–1889); "Truman Henry Safford," *Observatory: A Monthly Review of*

*Astronomy* 24 (1901): 307–309; and Arthur Searle, "Truman Henry Safford," *Proceedings of the American Academy of Arts and Sciences* 37 (1902): 654–656.

77 The analysis follows Bruno Latour, *Science in Action: How to Follow Scientists and Engineers through Society* (Cambridge, Mass.: Harvard University Press, 1987), 215–257.

78 See definition 1e in *OED Online*, s.v. "performance, n.," last modified June 2019, https://www.oed.com/view/Entry/140783.

79 See Beadie, "Student Markets."

80 Theodore M. Porter, *Trust in Numbers: The Pursuit of Objectivity in Science and Public Life* (Princeton: Princeton University Press, 1995); and Beadie, "Student Markets."

81 See Committee of Ten scholarship, including Theodore R. Sizer, *Secondary Schools at the Turn of the Century* (New Haven, Conn.: Yale University Press, 1964); David Lindsay Roberts, *American Mathematicians as Educators, 1893–1923: Historical Roots of the "Math Wars"* (Boston: Docent, 2012); and James K. Bidwell and Robert G. Clason, eds., *Readings in the History of Mathematics Education* (Washington, D.C.: National Council of Teachers of Mathematics, 1970), 129–141.

82 National Educational Association, "Supt. J. M. Greenwood," in *Journal of Proceedings and Addresses, Session of the Year 1894, Held at Asbury Park, New Jersey* (St. Paul, Minn.: Pioneer, 1895), 454.

83 Edwin P. Seaver, *Fourteenth Annual Report of the Superintendent of Public Schools of the City of Boston* (March 1894), later published in the *Annual Report of the School Committee of the City of Boston 1894* (Boston: Rockwell and Churchill, 1894), 11.

84 "Supt. J. M. Greenwood," in "Discussion" of W. T. Harris's "The Curriculum for Secondary Schools," in National Educational Association, *Journal of Proceedings . . . of the Year 1894*, 514–515.

85 Joseph M. Rice, "Educational Research," *Forum* 34 (July 1902): 118; and Joseph M. Rice, "Educational Research: A Test in Arithmetic," *Forum* 34 (October 1902): 283–297. Also, see Kidwell, Ackerberg-Hastings, and Roberts, *American Mathematics Teaching*, 39–41.

86 Joseph M. Rice, "Educational Research: Causes of Success and Failure in Arithmetic," *Forum* 34 (January 1903): 437–452.

87 E. L. Thorndike, "Animal Intelligence: An Experimental Study of the Associative Processes in Animals," *Psychological Review Monograph Supplement* 2 (1898). Also, see Geraldine Joncich, *The Sane Positivist: A Biography of Edward L. Thorndike* (Middletown, Conn.: Wesleyan University Press, 1968); and Paul Chance, "Thorndike's Puzzle Boxes and the Origins of the Experimental Analysis of Behavior," *Journal of the Experimental Analysis of Behavior* 72 (1999): 433–440.

88 Edward L. Thorndike, *Educational Psychology* (New York: Lemcke and Buecher, 1903).

89 Kidwell, Ackerberg-Hastings, and Roberts, *American Mathematics Teaching*, 41.

90 Kidwell, Ackerberg-Hastings, and Roberts, 41–43.

91 Quoted in Kidwell, Ackerberg-Hastings, and Roberts, 41–43.

92 See Ervin V. Johanningmeier, "The Transformation of Stuart Appleton Courtis: Test Maker and Progressive," *American Educational History Journal* 31, no. 2 (2004): 202–210; and Kidwell, Ackerberg-Hastings, and Roberts, *American Mathematics Teaching*, 43–45.

93 S. A. Courtis, "Our School as a Contributor to Educational Progress," *Rivista* (Detroit: Liggett School, 1910). As quoted in Kidwell, Ackerberg-Hastings, and Roberts, *American Mathematics Teaching*, 43.

94 See figure 3.1 in Kidwell, Ackerberg-Hastings, and Roberts, *American Mathematics Teaching*, 44.

95 Kidwell, Ackerberg-Hastings, and Roberts, 45–50. Also, see Johanningmeier, "Stuart Appleton Courtis."

96 Kidwell, Ackerberg-Hastings, and Roberts, *American Mathematics Teaching*, 50–52.

97 Agnes T. Rogers, "Tests of Mathematical Ability—Their Scope and Significance," *Mathematics Teacher* 11, no. 4 (June 1919): 156–157.

98 David Eugene Smith, "On Improving Algebra Tests," *Teachers College Record* 25, no. 2 (March 1923): 87–88.

99 Krug, *Shaping*, 89.

100 It seems that the question was actually intimately connected to many of the aims of the Committee of Ten. When states developed their own initiatives based on the NEA's proceedings, they considered the same question, too, albeit on a regional level. See, for instance, R. W. Jones, "Our Proposed New Requirements for Admission to College," *School Review* 9, no. 2 (1901): 112–114.

101 Krug, *Shaping*, 146–149.

102 Nicholas Murray Butler, "Uniform College Entrance Requirements with a Common Board of Examiners," in *Proceedings of the 13th Annual Convention of the Association of Colleges and Preparatory Schools of the Middle States and Maryland, Held at State Normal School, Trenton N.J., Friday and Saturday 1–2 December 1899* (Albany: University of the State of New York Press, 1900), 43–49; and Charles W. Eliot, "Discussion" to Butler's "Uniform College Entrance Requirements," in *Proceedings of . . . the Association of Colleges and Preparatory Schools of the Middle States and Maryland . . . 1899*, 85–86.

103 College Entrance Examination Board of the Middle States and Maryland, *First Annual Report of the Secretary* (New York: The Board, 1901), 1.

104 Kidwell, Ackerberg-Hastings, and Roberts, *American Mathematics Teaching*, 50–52.

105 Kidwell, Ackerberg-Hastings, and Roberts, 44.

106 See Nicholas Lemann, *The Big Test: The Secret History of the American Meritocracy* (New York: Farrar, Straus and Giroux, 1999).

107 Sister Mary Fides Gough, "Why Failures in Mathematics? Mathemaphobia: Causes and Treatments," *Clearing House: A Journal of Educational Strategies, Issues and Ideas* 28, no. 5 (1954): 290–294.

108 Mark H. Ashcraft, "Math Anxiety: Personal, Educational, and Cognitive Consequences," *Current Directions in Psychological Science* 11, no. 5 (2002): 181–185.

109 Frank C. Richardson and Richard M. Suinn, "The Mathematics Anxiety Rating Scale," *Journal of Counseling Psychology* 19 (1972): 551–554.

110 Ian M. Lyons and Sian L. Beilock, "When Math Hurts: Math Anxiety Predicts Pain Network Activation in Anticipation of Doing Math," *PLoS One* 7, no. 10 (2012): e48076. Also, see Sian L. Beilock, "Math Performance in Stressful Situations," *Current Directions in Psychological Sciences* 17 (2008): 339–343; Sian L. Beilock, *Choke: What the Secrets of the Brain Reveal about Getting It Right When You Have To* (New York: Simon and Schuster, 2010); Erin A. Maloney and Sian L. Beilock, "Math Anxiety: Who Has It, Why It Develops, and How to Guard against It," *Trends in Cognitive Sciences* 16, no. 8 (2012): 404–406; and Ian M. Lyons and Sian L. Beilock,

"Mathematics Anxiety: Separating the Math from the Anxiety," *Cerebral Cortex* 22 (2012): 2102–2110.

111 Ray Hembree, "The Nature, Effects, and Relief of Mathematics Anxiety," *Journal for Research in Mathematics Education* 21, no. 1 (1990): 33–46.

112 Gary Scarpello, "Helping Students Get past Math Anxiety," *Techniques: Connecting Education and Careers* 82, no. 6 (2007): 34–35.

113 National Council of Teachers of Mathematics, *Curriculum and Evaluation Standards for School Mathematics* (Reston, Va.: NCTM, 1989); and National Council of Teachers of Mathematics, *Mathematics Anxiety, Supplemental* (Reston, Va.: NCTM, 1995).

114 Marilyn Curtain-Phillips, *Math Attack: How to Reduce Math Anxiety in the Classroom, at Work and in Everyday Personal Use* (self-pub., 1999); and Marilyn Curtain-Phillips, "Math Anxiety," *Marilyn Curtain-Phillips*, last modified July 2016, https://www.marilyncurtainphillips.com.

## Conclusion

1 Andrew Hacker, *The Math Myth and Other STEM Delusions* (New York: New Press, 2016).

2 When I helped organize a workshop on STEAM in 2016, we represented the instrumental perspective with Helen Small, *The Value of the Humanities* (Oxford: Oxford University Press, 2013), 59–88; and J. Bradford Hipps, "To Write Better Code, Read Virginia Woolf," *New York Times*, 22 May 2016, SR7.

3 Nicolas Rose, "What Positive Psychology Can Learn from the Humanities," *Mappalicious: The German Side of Positive Psychology*, posted 18 June 2014, https://mappalicious.com/2014/06/18; and Mirabai Bush, "Mindfulness in Higher Education," *Contemporary Buddhism* 12, no. 1 (2011): 183–197.

4 Bruno Latour, "Morality and Technology: The End of the Means," *Theory, Culture and Society* 19 (2002): 247–260; and Holland Cotter, "Placement Is Politics in Brooklyn Museum Reinstallation," *New York Times*, 20 May 2016, C19.

5 Margaret A. Nash, Danielle C. Mireles, and Amanda Scott-Williams, "'Mattie Matix' and Prodigal Princes: A Brief History of Drag on College Campuses from the Nineteenth Century to the 1940s," in *Rethinking Campus Life: New Perspectives on the History of College Students in the United States*, ed. Christine A. Ogren and Marc A. VanOverbeke (New York: Palgrave Macmillan, 2018), 61–90.

6 There is an especially detailed description in Alexandra Robbins, *Secrets of the Tomb: Skull and Bones, the Ivy League, and the Hidden Paths of Power* (Boston: Little, Brown, 2002), 31–33.

7 Stephen Harris et al., "17: Science by the Seaside," 13 September 2017, in *Anthill*, produced by The Conversation UK, podcast, MP3 audio, 35:40, accessed 8 September 2018, http://theconversation.com/anthill-17-science-by-the-seaside-84008. For the advertising of the event, see "Theatre Events 2017," *University of Brighton Arts and Culture*, posted 8 September 2017, http://arts.brighton.ac.uk/whats-on/sallis-benney-events/theatre-events-2017/sep-2017/british-science-festival-at-the-university-of-brighton4; and "British Science Festival 2017," *London Mathematical Society Newsletter* 471 (July 2017): 29.

8 Jerry P. King, *The Art of Mathematics* (Mineola, N.Y.: Dover, 1992), especially 135–137.

9 See JoAnne S. Growney, "Mathematics and the Arts—a Bibliography," *Humanistic Mathematics Network Journal* 8 (1993): 24–36.

10 John McCleary and Audrey McKinney, "What Mathematics Isn't," *Mathematical Intelligencer* 8, no. 3 (September 1986): 51–53.

11 Thomas S. Kuhn, *The Structure of Scientific Revolutions* (Chicago: University of Chicago Press, 1970 [1962]), 92–110.

12 McCleary and McKinney, "What Mathematics Isn't," 51–53.

13 A highly cited recent example is President's Council of Advisors on Science and Technology, *Engage to Excel: Producing One Million Additional College Graduates with Degrees in Science, Technology, Engineering, and Mathematics* (Washington, D.C.: PCAST, 2012).

14 Hacker, *Math Myth*, 11.

15 Jean Webster, *When Patty Went to College* (New York: Century, 1903), chap. 13; Charles E. Bolton, *The Harris-Ingram Experiment* (Cleveland: Burrow Brothers, 1905), chap. 10; and Julia A. Schwartz, *Elinor's College Career* (Boston: Little, Brown, 1906), 95–101.

16 Helen Lee Sherwood, "Informal Dramatics," in *The Vassar Miscellany: Vassar, 1865–1915, from the Undergraduate Point of View, Fiftieth Anniversary Number* (Poughkeepsie, N.Y.: Vassar College, 1915), 111. For historical views of theater as "corrupting," see John F. Kasson, *Rudeness and Civility: Manners in Nineteenth-Century Urban America* (New York: Hill & Wang, 1990), 239–246.

17 Alan Alda, *If I Understood You, Would I Have This Look on My Face? My Adventures in the Art and Science of Relating and Communicating* (New York: Random House, 2016), 139–151.

18 Tom Kelly, *Ten Faces of Innovation* (London: Profile, 2006); and Michael Lewrick, Patrick Link, and Larry Leifer, *The Design Thinking Playbook* (Hoboken, N.J.: Wiley, 2018).

19 N. W. Taylor Root, *School Amusements; or, How to Make the School Interesting* (New York: A. S. Barnes, 1857). Emphasis added.

20 See Kay Harel, "When Darwin Flopped: The Rejection of Sexual Section," *Sexuality and Culture* 5 (Autumn 2001): 29–42; Mike Hawkins, *Social Darwinism in European and American Thought, 1860–1945* (Cambridge: Cambridge University Press, 1997); Sally Gregory Kohlstedt and Mark R. Jorgensen, "'The Irrepressible Woman Question': Women's Responses to Evolutionary Ideology," in *Disseminating Darwinism: The Role of Place, Race, Religion, and Gender*, ed. Ronald Numbers and J. Stenhouse (Cambridge: Cambridge University Press, 1999), 267–293; and Kimberly A. Hamlin, *From Eve to Evolution: Darwin, Science, and Women's Rights in Gilded Age America* (Chicago: University of Chicago Press, 2014), 128–130.

21 See Lynn D. Gordon, *Gender and Higher Education in the Progressive Era* (New Haven, Conn.: Yale University Press, 1990), 52–84; and David M. Stameshkin, *The Town's College: Middlebury College, 1800–1915* (Middlebury, Vt.: Middlebury College, 1985), 193–227.

22 Wilbur H. Dutton, "Measuring Attitudes toward Arithmetic," *Elementary School Journal* 55, no. 1 (September 1954): 24–31.

23 Dutton, 28.

24 Culminating in Lewis R. Aiken, "Two Scales of Attitude toward Mathematics," *Journal for Research in Mathematics Education* 5, no. 2 (March 1974): 61–71.

25 Elizabeth Fennema and Julia Sherman, "Fennema-Sherman Mathematics Attitudes Scales: Instruments Designed to Measure Attitudes towards the Learning of Mathematics by Females and Males," *Journal for Research in Mathematics Education* 7, no. 5 (November 1976): 324–326. Also, see "Elizabeth Fennema," in *Notable Women*

*in Mathematics, a Biographical Dictionary*, ed. Charlene Morrow and Teri Perl (Westport, Conn.: Greenwood, 1998), 51–56; and "Elizabeth Hammer Fennema," in *Who's Who of American Women*, 7th ed. (Wilmette, Ill.: Marquis Who's Who, 1973).

26  Fennema and Sherman, "Fennema-Sherman Mathematics."

27  Fennema and Sherman.

28  Janet Melancon, Bruce Thompson, and Shirley Becnel, "Measurement Integrity of Scores from the Fennema-Sherman Mathematics Attitudes Scales: The Attitudes of Public School Teachers," *Educational and Psychological Measurement* 54, no. 1 (March 1994): 187–192.

29  Fiona Mulhern and Gordon Rae, "Development of a Shortened Form of the Fennema-Sherman Mathematics Attitudes Scales," *Educational and Psychological Measurement* 58, no. 2 (April 1998): 295–306; Martha Tapia and George Marsh II, "An Instrument to Measure Mathematics Attitudes," *Academic Exchange Quarterly* 8, no. 2 (Summer 2004): 16; and Martha Tapia and George Marsh II, "The Relationship of Math Anxiety and Gender," *Academic Exchange Quarterly* 8, no. 2 (Summer 2004).

30  Tapia and Marsh, "Math Anxiety and Gender."

31  Paul Ernest, "Why Teach Mathematics? The Justification Problem in Mathematics Education," in *Justification and Enrolment Problems in Education Involving Mathematics or Physics*, ed. J. H. Jensen, M. Niss, and T. Wedege (Roskilde, Den.: Roskilde University Press, 1998), 33–55.

32  Elizabeth A. Flynn, "Feminism and Scientism," *College Composition and Communication* 46, no. 3 (October 1995): 353–368.

33  [United States] National Gallery of Art, *Highlights from the National Gallery of Art* (Washington, D.C.: National Gallery of Art, 2016), 233.

34  For instance "Winslow Homer in the National Gallery of Art," *National Gallery of Art*, last modified 2019, https://www.nga.gov/features/slideshows/winslow-homer-in-the-national-gallery-of-art.html.

35  "Winslow Homer."

36  Judith Butler, *Gender Trouble: Feminism and the Subversion of Identity* (New York: Routledge, 2008 [1990]), 34.

37  Ian M. Lyons and Sian L. Beilock, "When Math Hurts: Math Anxiety Predicts Pain Network Activation in Anticipation of Doing Math," *PLoS One* 7, no. 10 (2012): e48076. Also, see Sian L. Beilock, "Math Performance in Stressful Situations," *Current Directions in Psychological Sciences* 17 (2008): 339–343; Sian L. Beilock, *Choke: What the Secrets of the Brain Reveal about Getting It Right When You Have To* (New York: Simon and Schuster, 2010); Erin A. Maloney and Sian L. Beilock, "Math Anxiety: Who Has It, Why It Develops, and How to Guard against It," *Trends in Cognitive Sciences* 16, no. 8 (2012): 404–406; and Ian M. Lyons and Sian L. Beilock, "Mathematics Anxiety: Separating the Math from the Anxiety," *Cerebral Cortex* 22 (2012): 2102–2110.

38  McCleary and McKinney, "What Mathematics Isn't," 51–53.

39  The sociologist Herbert Mehrtens has provided a detailed bibliography of the relevant studies that consider the social embeddedness of mathematics as well as some key detractors: Herbert Mehrtens, "Social History of Mathematics" and "Select Bibliography," in *Social History of Nineteenth Century Mathematics*, ed. Herbert Mehrtens, Henk Bos, and Ivo Schneider (Boston: Birkhäuser, 1981), 257–299. Also see H. J. M. Bos, *Lectures in the History of Mathematics* (Providence, R.I.: American

Mathematical Society, 1993); Sal Restivo, *Mathematics in Society and History: Sociological Inquiries* (Dordrecht, Neth.: Kluwer Academic, 1992); and Karen Parshall and Adrian C. Rice, "The Evolution of an International Mathematical Research Community, 1800–1945: An Overview and an Agenda," in *Mathematics Unbound: The Evolution of an International Mathematics Research Community, 1800–1945* (Providence, R.I.: American Mathematical Society, 2002).

40  "The Alda Center," *Alan Alda Center for Communicating Science*, last modified 2019, https://www.aldacenter.org/get-started/about-us.

41  Alda, *If I Understood.*

42  Hadassah Damien, "Ten Things Theater Taught Me That Are Useful for Human-Centered Design Facilitation," *UX Collective*, 4 January 2019, https://uxdesign.cc/b3c1036b5925.

43  For the neglect of math, see Karen Hunger Parshall and David E. Rowe, *The Emergence of the American Mathematical Research Community, 1876–1900: J. J. Sylvester, Felix Klein, and E. H. Moore* (Providence, R.I.: American Mathematical Society, 1994), x–xii; and Ivor Grattan-Guinness, "Does History of Science Treat of the History of Science? The Case of Mathematics," *History of Science* 28 (1990): 149–173. They point back to George Sarton, *The Study of the History of Mathematics* (New York: Dover, 1957 [1937]), 7.

# Index

Page numbers in *italics* refer to figures.

Wood, Frances, 107
Woody, Thomas, 93–94
World War I, 137
written testing, 14, 115–118, 123–124, 126; debates about, 131–133; math anxiety and, 139, 140, 141, 144; performance and, 133–139, 144

*Yale Literary Magazine*, 88
Yale math, 22, 25–27, 83

"Yale Report," 77, 78–81, 82, 85
Yale University, New Haven, Connecticut, 8, 15–16, 81, 145; curriculum, 77, 78; influence of, 96; mathematics at, 22–23, 25–27, 28, 43; relation to other colleges, 49, 80, 85; reputation of, 27; rules of, 82, 84; school traditions, 40–45, 74, 75; student demographics, 18. *See also* Burial of Euclid; Conic Sections Rebellion
youth cultures, 16

## About the Author

ANDREW FISS is an assistant professor in technical communication at Michigan Technological University. He has published in journals devoted to education, U.S. history, rhetoric, and STEAM. Mainly archival, his research also has led to a few outreach events connecting to current educational practices in the United States and the United Kingdom.

Printed in the United States
By Bookmasters